Table of Contents

1 Introduction to HVAC

Heating, ventilation, and air conditioning—HVAC—systems maintain comfortable indoor environments. They're crucial for building occupant well-being and productivity, especially considering the wide range of climates worldwide. Understanding HVAC fundamentals is vital for anyone working in building design, construction, or maintenance.

The core function of an HVAC system is straightforward: it controls temperature, humidity, and air quality. This involves managing the movement of air and heat, often through a complex network of components. Think of it like a sophisticated circulatory system for a building, constantly working to maintain a healthy internal climate. We'll explore this 'circulation' in detail throughout this chapter.

Consider a simple residential system. It might consist of a furnace for heating, an air conditioner for cooling, and a blower to circulate air. These components work together, usually governed by a thermostat, to respond to changing environmental conditions. But modern systems are far more intricate, integrating advanced technologies like variable-speed motors and smart controls. These advancements are designed to improve energy efficiency and enhance user control.

Different buildings have dramatically different HVAC needs. A large office complex will require a far more powerful and complex system than a small residential home. Similarly, a hospital with stringent air quality standards will have very specific HVAC requirements, potentially involving specialized filtration and air handling units to remove contaminants. These variances emphasize the need for careful system design tailored to the specific application.

HVAC system types vary based on several factors, such as building size, budget, and desired climate control level. Central systems, the most common, distribute conditioned air throughout a building via ducts. These systems can be highly efficient and provide uniform climate control, though installation can be complex. In contrast, split systems, typically found in residences and smaller commercial spaces, have separate indoor and outdoor units. These often offer a more cost-effective solution for smaller applications, needing less extensive ductwork. The choice depends entirely on the needs of the project at hand.

Heat transfer is central to HVAC. This involves three main processes: conduction, convection, and radiation. Conduction is the transfer of heat through direct contact, like heat flowing from a hot radiator to the surrounding air. Convection involves heat transfer through the movement of fluids, like hot air rising and cooler air sinking, creating air currents that help distribute the temperature more evenly. Radiation involves the emission of electromagnetic waves, like the warmth you feel from the sun or a heated surface. Understanding these processes is essential for designing efficient HVAC systems.

Refrigeration cycles are the heart of cooling systems. These cycles typically utilize refrigerants, which absorb heat from one area and release it in another. This principle is based on the thermodynamic properties of refrigerants, which transition between liquid and gas phases based on temperature and pressure. These processes happen in a vapor-compression cycle, a cycle consisting of a compressor, condenser, expansion valve, and evaporator. The careful engineering of these components ensures efficient heat transfer and appropriate cooling.

Air distribution systems are equally critical, ensuring that conditioned air reaches all areas of a building effectively. This involves designing and installing ductwork to deliver the air. Proper ductwork design is critical to avoid pressure drops and ensure uniform air distribution. There are considerations about the material selection and duct size, alongside air flow rates and pressure calculations. Efficient ductwork leads to improved system performance.

Proper system design involves understanding psychrometrics, the science of moist air. It considers factors such as temperature, humidity, and pressure, and how these parameters affect the comfort and energy efficiency of a building. Understanding psychrometric charts and calculations is essential for accurate HVAC design calculations. It is a critical tool in HVAC design and system performance. A good grasp of psychrometrics is crucial for a proficient HVAC professional.

In conclusion, a foundational understanding of HVAC principles encompasses heat transfer mechanisms, thermodynamic cycles, refrigeration cycles, air distribution, and psychrometrics. This knowledge forms the basis for effective system design, allowing for the creation of efficient, comfortable, and sustainable building environments. Mastering these fundamentals is a crucial first step towards becoming proficient in the field of HVAC engineering.

1.1 Basic HVAC Principles

Understanding the fundamental principles of heating, ventilation, and air conditioning (HVAC) is crucial for any professional in the field. This involves grasping core concepts that govern how these systems function effectively, influencing everything from comfort levels to energy efficiency. A strong foundation in these basics allows for a more thorough understanding of advanced concepts and problem-solving later on. Consider it the bedrock upon which all other HVAC knowledge is built.

Heat transfer, a primary component of HVAC, operates on three key mechanisms: conduction, convection, and radiation. Conduction involves the transfer of heat through direct contact, like the heat traveling through the walls of a building. Convection relies on the movement of fluids, such as air or water, to distribute heat. Imagine how a radiator warms a room: the hot radiator heats the air around it, which then circulates, distributing the heat. Lastly, radiation involves the emission of electromagnetic waves, a process that's evident in how the sun's warmth reaches the earth. Understanding these mechanisms helps in designing systems that efficiently transfer heat where needed and prevent heat loss.

Air movement is another essential principle. HVAC systems manipulate airflow to distribute conditioned air effectively throughout a building. Factors like pressure differences, fan design, and ductwork configuration influence the speed and direction of airflow. Proper airflow ensures uniform temperature and humidity levels, leading to comfortable indoor environments. Inadequate airflow, however, can result in hot and cold spots, leading to discomfort and inefficient energy use. Think of it as the circulatory system of your building, carrying conditioned air to every corner. The design of this system is critical for its effectiveness.

The concept of sensible and latent heat significantly impacts HVAC design. Sensible heat is the heat we can feel; it's the energy that alters the temperature of a substance without changing its state. Latent heat, on the other hand, is the energy required to change a substance's state, such as the energy needed to evaporate water. Understanding these differences is critical for calculating cooling loads and choosing the right HVAC equipment. Consider the need for dehumidification in a humid climate: removing moisture from the air involves removing latent heat, distinct from simply lowering the temperature.

Psychrometrics provides the tools for analyzing moist air

properties, including temperature, humidity, and enthalpy. Understanding these properties is essential for correctly sizing HVAC equipment, especially for cooling systems. A psychrometric chart provides a visual representation of these relationships, enabling engineers to accurately calculate cooling and dehumidification requirements for specific environments. This data becomes crucial when evaluating system efficiency and selecting suitable refrigerants. It allows for a precise tailoring of the system to specific climate and building needs.

Finally, thermodynamics plays a vital role in understanding the energy transfer processes within HVAC systems. Understanding concepts such as the first and second laws of thermodynamics helps in optimizing system efficiency and minimizing energy consumption. This knowledge influences the design and selection of components, such as compressors, expansion valves, and heat exchangers, all of which impact system efficiency. The goal is to maximize the desired heat or cold transfer while minimizing wasted energy. Proper understanding of thermodynamics is what bridges the practical application with theoretical physics.

In short, a solid understanding of these basic principles – heat transfer mechanisms, airflow dynamics, sensible and latent heat, psychrometrics, and thermodynamics – is the foundation for successful HVAC design and operation. These fundamental concepts underpin more advanced topics, forming the groundwork for effective problem-solving and optimizing the efficiency and performance of any HVAC system. Ignoring these principles leads to ineffective designs and costly mistakes. Mastery of these principles is essential for becoming a proficient HVAC professional.

1.2 HVAC System Types

Many HVAC systems exist, each suited to different building sizes, needs, and budgets. Understanding these differences is crucial for selecting the right system. Consider a small residential unit versus a large commercial building – the appropriate technology varies significantly. The homeowner might opt for a simple split system, while the skyscraper necessitates a much more complex, potentially multi-zone, solution. The fundamental principles, however, remain the same, involving heat transfer and air movement.

Split systems, commonly found in residences, consist of an outdoor condenser unit and an indoor evaporator unit, connected by refrigerant lines. They're relatively inexpensive, efficient for their size, and simple to install and maintain.

However, they might not be ideal for large spaces requiring zoned climate control or for buildings with unique architectural challenges. Their capacity limitations need careful consideration during the design phase. This system's simplicity makes it a cost-effective solution for many applications, but that simplicity also means limited flexibility. Accurate sizing is paramount to ensure optimal performance and avoid common issues like insufficient cooling or excessive energy consumption. Regular maintenance, including cleaning the coils, is essential for prolonging the lifespan and efficiency of the system. These systems are also particularly effective in warmer climates, while supplemental heating may be necessary in colder regions.

Packaged units are self-contained systems, often installed on rooftops or in mechanical rooms. These units combine the condenser, evaporator, and other components into a single package. This configuration simplifies installation, making them efficient choices for smaller commercial buildings or multi-family dwellings. They are advantageous for applications that need a compact, integrated solution. The all-in-one design reduces installation complexity compared to split systems; however, access for maintenance and repair should always be considered during the placement phase. The overall efficiency can vary significantly depending on the specific unit and its features. Energy efficiency ratings, such as SEER (Seasonal Energy Efficiency Ratio) are crucial aspects to consider, both for initial cost calculation and long-term operational expenses. Different models offer varying levels of capacity and features, to match specific requirements.

Central air conditioning systems utilize a central blower to circulate air through a network of ductwork. This type allows for climate control across multiple zones within a larger building, offering greater flexibility than packaged or split systems. These systems are the norm in larger commercial and residential buildings and require more significant initial investment because of the extended ductwork, and the need for more complex installation and balancing of the air distribution system. The possibility of zoned climate control is a significant advantage, enabling greater occupant comfort and energy savings by tailoring the temperature in each area. However, this complexity also brings potential downsides: duct leakage can lead to energy loss and uneven temperature distribution, necessitating regular inspections and maintenance. Additionally, proper ductwork design and installation are vital for efficient system operation. Improper sizing and design can lead to increased energy costs and reduced comfort. Consider the potential for noise transmission through the ductwork, especially in residential settings; careful material selection and installation practices are essential for noise control.

Heat pumps provide both heating and cooling using the same unit, reversing the refrigeration cycle to provide heat during colder months. This dual functionality makes them highly energy-efficient in regions with moderate climates, eliminating the need for separate heating and cooling systems. They offer significant energy savings when compared to systems that rely on separate heating and cooling technologies. The efficiency, however, can be impacted by extremely low temperatures, where supplemental heating might be required. Heat pumps come in various types, including air-source, ground-source (geothermal), and water-source, each with its unique installation requirements and cost implications. Geothermal heat pumps, for instance, offer exceptional energy efficiency but require substantial upfront investment for ground loop installation. Careful evaluation of climate conditions, energy costs, and the upfront cost is paramount before choosing this type of system. Proper maintenance is essential for keeping them running efficiently and safely.

Choosing the right HVAC system involves considering factors such as building size, climate, budget, and energy efficiency goals. Careful consideration of these variables, coupled with professional consultation, will ensure the selection of a system that meets the specific needs of the building and its occupants, optimizing both performance and long-term sustainability. This decision process should incorporate a life-cycle cost analysis, factoring in initial investment, operating costs, and maintenance requirements. This comprehensive evaluation enables informed decision-making, resulting in a more effective and energy-efficient HVAC system. Remember that ongoing maintenance is critical to the continued efficient and safe operation of any system. Neglecting these aspects can lead to significant energy waste, premature failure, and increased costs in the long run.

2 Fundamentals of Thermodynamics

Understanding the behavior of heat and its interaction with matter is fundamental to HVAC system design. Heat transfer, the process by which thermal energy moves from one region to another, occurs through three primary mechanisms: conduction, convection, and radiation. Conduction involves the transfer of heat through direct contact within a material, with materials possessing varying thermal conductivities influencing the rate of heat flow. Convection, on the other hand, relies on the movement of fluids (liquids or gases) to transfer heat; warmer fluids rise, carrying heat energy, while cooler fluids descend, creating a circulation pattern. This principle is pivotal in understanding how air circulates within HVAC systems, influencing temperature distribution and efficiency. Finally, radiation involves the transmission of heat energy through electromagnetic waves, a process unaffected by the presence or absence of a medium. Understanding the relative contributions of these three mechanisms in various components of an HVAC system is crucial for optimizing design and performance.

The efficiency of any HVAC system hinges heavily on its thermodynamic cycle. The Carnot cycle, while theoretical, serves as a benchmark for understanding maximum possible efficiency. It consists of two isothermal and two adiabatic processes, achieving its high efficiency through idealized conditions. In reality, however, HVAC systems utilize cycles that approximate the Carnot cycle but accommodate practical limitations. The Rankine cycle, commonly employed in power generation and some heating applications, involves the vaporization and condensation of a working fluid, driven by a heat source and sink. Understanding the variations and efficiencies of different cycles—such as variations in Rankine cycles—is essential for selecting appropriate systems and refrigerants for specific climate conditions and design requirements. Factors like pressure, temperature, and fluid properties significantly impact the thermodynamic performance, and careful analysis of these parameters informs effective system design.

Psychrometrics forms the crucial link between thermodynamics and the practical application in HVAC design. It deals with the thermodynamic properties of moist air—a mixture of dry air and water vapor. Understanding psychrometric charts and calculations is paramount for determining the properties of air at various points within an HVAC system, such as temperature, humidity, and enthalpy. These properties, in turn, influence the system's ability to control the environment's comfort levels. For instance, knowing the dew point temperature helps predict the potential for condensation, a critical factor in preventing moisture-related problems in buildings. Calculations involving specific humidity, relative humidity, and enthalpy, often performed using psychrometric charts, allow designers to choose optimal equipment sizing and operating conditions to meet the building's needs. Accurate psychrometric calculations ensure proper cooling and dehumidification, optimizing the HVAC system for energy efficiency and occupant comfort.

Let's consider a practical example. Suppose you're designing an HVAC system for a large office building in a humid climate. You need to carefully consider the heat load, which encompasses heat gain from external sources like solar radiation and internal sources such as lighting and occupancy. Through conduction calculations, you determine the heat transfer through the building's walls and windows. Convection calculations help estimate the heat transfer due to air infiltration and the movement of air within the building. Radiation calculations are crucial to determining the amount of heat gained from sunlight passing through the windows. Using psychrometrics, you can then calculate the required cooling capacity to maintain comfortable indoor temperatures and humidity levels. The choice of refrigeration cycle and refrigerant type must align with these calculations. You can then employ energy-efficient measures by optimizing the cooling cycle, using high-efficiency equipment, and controlling air infiltration. Finally, a thorough understanding of thermodynamics and psychrometrics helps ensure optimal system performance and efficient use of resources. Ignoring these principles can lead to oversized equipment, excessive energy consumption, and suboptimal indoor conditions, resulting in increased operational costs and reduced comfort.

Choosing the right refrigerant is pivotal in HVAC system efficiency. Refrigerants exhibit different thermodynamic properties, impacting the cycle's performance and environmental impact. Factors like their boiling and condensing points, along with their specific heat capacity, influence the system's efficiency and cooling capacity. Additionally, the global warming potential (GWP) and ozone depletion potential (ODP) of various refrigerants are critical considerations in today's environmentally conscious designs, driving the selection of low-GWP and zero-ODP refrigerants. The shift towards refrigerants with minimal environmental impact necessitates a thorough understanding of their properties and the regulations governing their use. This careful selection process, informed by a grasp of thermodynamic principles and environmental regulations, ensures the long-term sustainability and responsible operation of HVAC systems. Moreover, selecting environmentally friendly refrigerants contributes to the wider goal of reducing the carbon footprint associated with building operations, aligning with broader sustainability initiatives. The long-term cost savings associated with energy efficiency and the avoidance of environmental penalties makes this careful evaluation a cornerstone of responsible HVAC system design.

2.1 Heat Transfer Mechanisms

Heat is transferred through three primary mechanisms: conduction, convection, and radiation. Conduction involves the transfer of heat through direct contact between molecules, with higher energy molecules transferring energy to lower energy neighbors. Think of a hot pan warming your hand—the heat moves directly from the pan's surface through your skin. The rate of conductive heat transfer depends heavily on the material's thermal conductivity; metals conduct heat exceptionally well, while insulators, like fiberglass, do poorly. Understanding these properties is crucial when selecting materials for ductwork, insulation, and building envelopes. Effective insulation minimizes conductive heat loss in winter and heat gain in summer, optimizing HVAC system performance and energy efficiency. Incorrect material choices can lead to significant energy waste.

Convection, unlike conduction, involves heat transfer through the movement of fluids—liquids or gases. Warm air, less dense than cooler air, rises, creating a natural convective current. This is how a radiator heats a room; the warm air from the radiator rises, circulating heat throughout the space. Forced convection, such as that achieved with fans and blowers in HVAC systems, accelerates this process, improving the distribution of heated or cooled air. Proper airflow design is paramount for efficient convection-driven heat transfer in HVAC systems. Inadequate air circulation can result in temperature stratification, where significant temperature differences exist between different zones within a room, leading to discomfort and inefficiency.

Radiation, on the other hand, is the transfer of heat through electromagnetic waves. Unlike conduction and convection, it doesn't require a medium for transmission—heat can radiate through a vacuum. The sun warming the Earth is a prime example of radiant heat transfer. In HVAC, radiant heating and cooling systems utilize this mechanism, employing radiant panels or floors to directly heat or cool the occupants and objects in a space, rather than solely heating or cooling the air. These systems often prove more comfortable and energy-efficient than traditional convection-based systems, as they directly heat or cool surfaces rather than relying on air circulation. However, careful consideration of surface temperatures and material properties is required to avoid overheating or discomfort.

The relative importance of each mechanism varies depending on the specific situation. In a well-insulated building, radiation may play a more significant role in heat transfer, especially from sunlight penetrating windows. Conversely, in a poorly insulated building, conductive and convective heat losses will be more dominant. Effective HVAC system design must account for all three mechanisms to optimize energy efficiency and occupant comfort. For instance, minimizing conductive heat loss through proper insulation and airtight construction is a fundamental step towards reducing the load on the heating and cooling systems.

Consider a typical building's heat transfer dynamics. During summer, solar radiation through windows contributes substantially to the building's heat gain. This heat is then transferred through conduction to the interior surfaces, leading to increased indoor temperatures. Convection through air infiltration further contributes to the cooling load. Efficient HVAC design minimizes these unwanted heat transfers. Employing high-performance glazing to reduce solar heat gain, installing insulation to reduce conductive losses, and sealing air leaks to limit infiltration all reduce the need for excessive cooling.

Conversely, during winter, the primary heat losses are through conduction and convection. Heat escapes through poorly insulated walls and roofs, and air infiltration carries away warmed air. Minimizing these losses through careful building construction and insulation selection, is crucial. Additionally, understanding radiation helps in strategic placement of heat sources to maximize comfort. For example, radiant floor heating systems can provide even and comfortable warmth, minimizing the energy required compared to systems relying solely on convective air heating. An effective HVAC system must work in harmony with the building envelope, strategically mitigating heat transfer to maintain optimal indoor conditions.

Understanding heat transfer mechanisms is essential for effective HVAC system design. Failure to account for these mechanisms can result in undersized or oversized equipment, leading to increased energy consumption, thermal discomfort, and potential equipment failures. This understanding helps in the selection of appropriate materials, insulation levels, and system configurations to optimize performance and efficiency. It is not merely about selecting the right equipment but understanding how the system interacts with the building envelope and the environment. This is why a comprehensive understanding of thermodynamics and heat transfer is crucial for any professional working in the HVAC field. Therefore, grasping the intricacies of conduction, convection, and radiation is a fundamental prerequisite for mastering HVAC design and ensuring optimal system performance.

2.2 Thermodynamic Cycles

Understanding thermodynamic cycles is fundamental to HVAC system design. These cycles govern the transfer of heat and work, dictating the efficiency and effectiveness of heating and cooling processes. A firm grasp of these principles allows for informed decisions regarding system selection, component sizing, and overall system optimization, leading to significant energy savings and improved performance. Without a proper understanding of these processes, designing efficient and reliable HVAC systems is simply impossible.

The Carnot cycle, while idealized and not directly implemented in real-world systems, serves as a crucial benchmark. Its theoretical efficiency sets an upper limit for any heat engine. Studying the Carnot cycle helps to understand the fundamental limitations of heat transfer and the thermodynamic principles governing energy conversion. This understanding is transferable and helps engineers to better evaluate the performance of real-world systems, identifying areas for potential improvement and optimization. By comparing real cycles to the Carnot cycle, we can quantify the inefficiencies in actual HVAC equipment and pinpoint areas that need improvement.

Real-world HVAC systems often utilize modified Rankine cycles for refrigeration. This cycle, while less efficient than the theoretical Carnot, is more practical and adaptable. It involves a compressor, condenser, expansion valve, and evaporator, each playing a vital role in the refrigeration process. Understanding the function and interaction of these components is critical for troubleshooting and maintenance, allowing professionals to effectively identify and resolve malfunctions within a refrigeration system. This requires detailed knowledge of the processes occurring in each component and the working fluid's behavior throughout the cycle.

The vapor-compression refrigeration cycle, a practical application of the Rankine cycle, dominates the field of HVAC cooling. It relies on the phase change properties of refrigerants to achieve cooling. The refrigerant absorbs heat as it evaporates, drawing heat from the space being cooled, and then releases this heat as it condenses within the condenser. An understanding of the thermodynamic properties of the refrigerant, such as its pressure-enthalpy relationship, is essential for efficient system design and operation. Improper selection of a refrigerant can result in significant efficiency losses.

Beyond refrigeration, thermodynamic cycles also apply to heating systems. Gas-fired heating systems, for example, often use a modified Brayton cycle. This cycle involves compression, heat addition, expansion, and heat rejection stages. Efficient combustion and proper heat exchange are critical in maximizing the efficiency of these systems. An understanding of the various aspects of the cycle, such as heat transfer within the heat exchangers and the combustion efficiency, is paramount in creating optimal heating systems. The design needs to consider the appropriate gas flow rates, combustion temperatures, and heat exchanger characteristics for optimal performance.

Analyzing the performance of thermodynamic cycles often involves considering entropy and exergy. Entropy represents the degree of disorder in a system, while exergy represents the maximum useful work that can be extracted from a system. Understanding these concepts enables more refined analyses of system efficiency and identifies opportunities for improvement. Analyzing entropy generation throughout the cycle can help pinpoint sources of irreversibilities, offering insights into areas needing optimization. Minimizing entropy generation leads to more efficient HVAC systems.

Furthermore, different factors influence the real-world performance of thermodynamic cycles. These factors include pressure losses in the system, heat transfer inefficiencies, and the frictional losses within components. These losses reduce the overall effectiveness of the cycle. The design process must incorporate considerations of these losses to model a cycle more accurately. Account for these losses during system design ensures realistic performance expectations.

Finally, consider the environmental impact when analyzing thermodynamic cycles. The selection of refrigerants, for example, must prioritize environmental friendliness. Using environmentally responsible refrigerants that reduce ozone depletion and global warming potential is critical to sustainable HVAC design. This includes investigating alternatives with lower global warming potentials and evaluating the lifecycle environmental impact of each component. The choice of refrigerant significantly impacts a system's overall environmental footprint.

Therefore, a comprehensive understanding of thermodynamic cycles is essential for HVAC professionals. From evaluating the efficiency of systems to troubleshooting problems and optimizing design, thermodynamic principles form the bedrock of effective HVAC engineering. A thorough knowledge enables the development of more efficient, sustainable, and environmentally responsible HVAC systems, maximizing performance while minimizing environmental impact and operating costs. This knowledge translates directly into better system design, reduced energy consumption, and overall improved building performance.

2.3 Psychrometrics

Understanding psychrometrics is fundamental to successful HVAC design. It's the science of moist air, a crucial element in any heating, ventilation, and air conditioning system. Without a grasp of psychrometric principles, accurate calculations for humidity control, sensible and latent heat loads, and overall system performance are impossible. Mastering this topic isn't merely about memorizing formulas; it's about understanding the relationships between air temperature, pressure, humidity, and enthalpy. This understanding will empower you to design more efficient and effective systems.

The properties of moist air are complex, influencing everything from human comfort to equipment efficiency. Imagine a typical summer day; high temperatures and high humidity create a stifling environment, making people uncomfortable. A psychrometric chart visually represents these interactions, allowing you to quickly see how changes in temperature and humidity affect other properties like specific volume, enthalpy, and relative humidity. It's a powerful tool that simplifies a seemingly complex problem. You'll quickly find it becomes invaluable during design and troubleshooting.

Let's delve into some key psychrometric properties. Dry-bulb temperature is the air temperature measured by a standard thermometer. Wet-bulb temperature, measured using a thermometer with a wet wick, accounts for evaporative cooling—a critical factor in humidity calculations. Dew point temperature is the temperature at which condensation begins. Understanding these three temperatures helps you determine the amount of moisture in the air and is critical in preventing condensation issues in your designs. Ignoring these values often leads to condensation on surfaces, resulting in issues like mold growth and structural damage.

Relative humidity, often expressed as a percentage, represents the ratio of the actual water vapor present in the air to the maximum amount the air can hold at a given temperature. It's a crucial parameter for human comfort; excessive humidity can cause discomfort, while very low humidity can dry out skin and mucous membranes. Absolute humidity, measured in grains of moisture per pound of dry air, provides a more direct measure of the moisture content. Choosing the appropriate metric is essential depending on the design objective. Calculations based on either relative or absolute humidity must be precise, otherwise, the entire system can underperform. Enthalpy, representing the total heat content of the air, is critical in calculating heating and cooling loads. It considers both sensible heat (related to temperature change) and latent heat (related to moisture changes during phase transitions). Accurate enthalpy calculations are essential for sizing equipment correctly. Underspecifying equipment leads to inadequate climate control, causing discomfort for occupants and potentially harming equipment lifespan. Over-specifying can waste energy and financial resources.

Understanding the psychrometric chart is critical for applying these principles effectively. The chart is a graphical representation of various air properties at different temperatures and humidity levels. It allows for quick estimations and visualizations of changes in air properties as a result of various processes, such as heating, cooling, humidification, and dehumidification. The chart is like a roadmap that guides your design decisions. Every HVAC engineer needs to feel comfortable interpreting a psychrometric chart to use it as an effective design tool.

Beyond the basic properties, understanding psychrometric processes is vital. The processes of sensible heating, sensible cooling, adiabatic mixing, and evaporative cooling all affect the state of the air, and each has implications for energy efficiency and system performance. These processes determine the temperature, humidity, and enthalpy changes during the HVAC system operation. The changes are not always predictable and require expertise to anticipate them during the design and execution phases of a project.

Therefore, carefully consider the impact of each process on the overall design and energy efficiency of your project. For instance, sensible heating increases the air temperature without changing its moisture content, while latent cooling removes moisture without significantly altering the temperature. Adiabatic mixing involves combining two air streams with different properties, and evaporative cooling uses water evaporation to reduce the air temperature. These factors are inextricably linked to the overall efficiency and performance of your systems.

Accurate psychrometric calculations are crucial for sizing HVAC equipment correctly. Using the psychrometric chart and relevant equations, you can determine the heating and cooling loads for a space, based on its occupancy, thermal loads, and desired comfort conditions. Incorrect calculations can lead to inefficient operations. For example, a system that is too small for the required load will fail to maintain the target conditions while an oversized system will waste energy. Finally, remember that this is not just about theoretical knowledge. Psychrometrics is a practical tool for HVAC professionals. From sizing equipment to troubleshooting system problems, a thorough understanding of moist air properties will consistently impact your work. It's a fundamental skill that separates proficient designers from truly exceptional ones. You will see the direct application of this knowledge every day in your work. Become fluent in the language of psychrometrics; it's the key to unlocking efficient and effective HVAC design.

3 Air Distribution Systems

Proper duct design is paramount. Undersized ducts restrict airflow, leading to inadequate heating or cooling, increased energy consumption, and potentially noisy operation. Conversely, oversized ducts waste energy by moving more air than necessary. Accurate sizing calculations, considering factors such as duct length, friction losses, and the required airflow rate for each zone, are crucial. Material selection also plays a significant role; galvanized steel is common for its durability and cost-effectiveness, while flexible ducts offer convenience in installations with limited space, but often come with higher friction losses. Remember to meticulously plan for access points for maintenance and cleaning; neglecting this can lead to costly repairs and system inefficiencies down the line. Careful consideration should be given to the insulation of the ductwork to minimize heat gain or loss, maintaining consistent air temperature and reducing energy expenditure.

Airflow and pressure are intrinsically linked within an air distribution system. Understanding these relationships is key to optimizing performance. Pressure drop, the reduction in air pressure as air flows through the ductwork, is directly influenced by duct size, shape, length, and internal surface roughness. Maintaining a balance between sufficient airflow and minimized pressure drop is essential for efficient energy use. Airflow velocity needs careful consideration; excessively high velocities can generate noise and increased pressure drop, whereas excessively low velocities can lead to stratification and uneven temperature distribution. To address these potential issues, employ strategies like carefully choosing fan size and static pressure, optimizing duct routing to reduce the overall length and number of bends, and ensuring smooth transitions between ducts of varying sizes. This is where accurate calculations and professional judgment are indispensable.

Ventilation strategies are crucial for maintaining healthy indoor air quality. This involves the controlled introduction of fresh outdoor air and the removal of stale or contaminated indoor air. Various strategies exist, such as natural ventilation, utilizing windows and stack effects, or mechanical ventilation systems, employing fans to move air. The specific choice depends on factors like building design, climate, and occupant density. For instance, a building in a humid climate may require a system that effectively dehumidifies incoming air. Similarly, a densely occupied office space will necessitate higher airflow rates to dilute pollutants and maintain acceptable indoor air quality. Codes and standards set minimum ventilation rates to ensure occupant health, which must be meticulously followed for compliance. Designing a system that meets these standards while maximizing energy efficiency often requires a multidisciplinary approach, combining engineering principles with insights from public health.

Balancing an air distribution system is a critical process. It aims to ensure that the desired airflow is delivered to each zone within the building. This involves measuring the actual airflow in each duct and adjusting dampers to achieve the designed flow rates. Imbalances can arise from various sources, such as inaccurate duct sizing, leaks, or obstructions. Utilizing specialized tools, such as airflow measuring devices and balancing dampers, is essential to ensure the precision needed for optimal system performance. A well-balanced system will deliver consistent temperature and air quality to all areas, maximizing occupant comfort and energy efficiency. Moreover, meticulous documentation of the balancing process is critical for future maintenance and troubleshooting. Without it, identifying and rectifying future imbalances becomes significantly more challenging and time-consuming.

Troubleshooting air distribution systems involves systematically investigating and resolving problems. Common issues include insufficient airflow, uneven temperature distribution, excessive noise, and high energy consumption. A methodical approach, starting with visual inspections to identify obvious problems such as leaks or obstructions, is always the first step. This might involve checking for blockages within ductwork, examining the fan performance, or verifying the proper function of dampers. More advanced diagnostics may involve employing pressure readings at various points in the system or using specialized instruments such as thermal cameras to detect heat leaks or air leaks. Accurate diagnosis frequently relies on understanding the interrelationship between various system components; a problem in one area might manifest as a symptom in another, requiring a comprehensive analysis. Maintaining detailed records throughout the troubleshooting process simplifies tracking progress and ensuring that the root cause of any issue is accurately addressed.

3.1 Duct Design Basics

Proper duct design is crucial for efficient and effective HVAC systems. Inadequate design can lead to uneven air distribution, reduced system performance, increased energy consumption, and even potential health issues due to poor air quality. Understanding the basics, therefore, is paramount for any HVAC professional.

The primary goal of duct design is to deliver conditioned air to the occupied spaces within a building efficiently and uniformly. This involves selecting the appropriate duct size, shape, and materials, and arranging the ductwork layout to minimize pressure losses and ensure proper airflow. Factors like the building's layout, the heating and cooling loads of individual rooms, and the desired airflow rates all play significant roles.

Accurate load calculations form the foundation of any successful duct design. This involves determining the heating and cooling demands of each zone within the building, considering factors such as climate, building insulation, window sizes, and occupancy levels. Software programs and industry standards provide guidance in these calculations.

Once the loads are determined, the duct system must be sized appropriately. Duct sizing involves calculating the required airflow rate for each zone and selecting duct sizes that minimize pressure drop while maintaining acceptable velocity. Several methods exist for sizing ducts, including the equal-friction method and the static-regain method, each with its own set of advantages and disadvantages. The choice depends on factors such as the system's complexity and the desired level of accuracy. Detailed calculations are essential, and professional HVAC design software significantly streamlines this. Remember to always check your calculations and account for any potential deviations.

The selection of duct materials is equally important. Factors to consider include the material's cost, durability, resistance to corrosion, and its ability to maintain air quality. Common duct materials include galvanized steel, aluminum, and flexible ductwork. Galvanized steel is durable and relatively inexpensive, making it a popular choice for many applications. Aluminum offers lighter weight and better corrosion resistance. Flexible ductwork is often used in situations where space is limited or for connecting to individual diffusers. Choosing the correct material for the environment is crucial, whether it's a humid or dry climate. This affects the system's longevity.

Beyond material selection, the actual layout of the ductwork significantly influences system performance. A well-designed duct system minimizes bends and turns, reducing pressure drop and improving airflow efficiency. Strategic placement of ductwork also helps to ensure even air distribution throughout the building, preventing cold spots or excessively hot areas.

Pressure drop calculations are paramount. Pressure losses occur throughout the duct system due to friction, bends, and other fittings. Accurate pressure drop calculations are essential to ensure that the system can deliver the required airflow rate to each zone. Using appropriate equations and factors from design guides assists in this calculation.

Furthermore, the integration of dampers and other control devices is crucial for efficient system operation. Dampers allow for adjusting airflow to individual zones, balancing the system, and meeting specific needs. These are integrated into the design to allow for future adjustments if necessary. Proper damper sizing and placement will minimize the likelihood of future problems. Again, using professional software and calculations are paramount.

Finally, thorough testing and balancing of the completed system are essential to ensure it performs as designed. This involves measuring airflow rates at each diffuser and making adjustments to the dampers to achieve the desired airflow distribution. This is done on-site after installation and plays a critical role in fine-tuning the overall system.

Careful attention to these details is vital for creating a high-performing, energy-efficient HVAC system. Understanding and applying these principles significantly reduces common installation errors and promotes a system operating at peak efficiency. While advanced software can assist, a thorough grasp of fundamental principles is essential for effective duct design. Ignoring these fundamentals leads to systems with problems.

3.2 Airflow and Pressure

Understanding airflow and pressure is fundamental to effective HVAC system design. Properly balanced airflow ensures even temperature distribution, preventing uncomfortable hot and cold spots within a building. Neglecting this crucial aspect can lead to significant energy waste and occupant dissatisfaction. A well-designed system, however, provides a comfortable and efficient environment, optimizing energy consumption and enhancing overall building performance. Consider the impact of pressure differences on system operation; these dictate the direction and volume of airflow throughout the ductwork.

Air pressure, measured in Pascals (Pa) or inches of water column (in. wg), is the force exerted by air per unit area. Positive pressure indicates that the air pressure inside a space is higher than the surrounding atmospheric pressure. Conversely, negative pressure means the internal pressure is lower. These pressure differences are responsible for driving airflow; air naturally moves from areas of higher pressure to areas of lower pressure. Think of it like water flowing downhill—air flows from high-pressure zones to low-pressure zones. This simple principle governs the movement of air throughout your entire HVAC system.

Designing for proper pressure relationships requires careful consideration of several factors. These include the size and configuration of the ductwork, the capacity of the air handling unit (AHU), and the number and location of supply and return registers. For instance, inadequate duct sizing can lead to excessive pressure drops, reducing airflow and impacting system performance. Conversely, over-sized ducts can cause uneven airflow distribution. Therefore, a balance is key. Precise calculations are needed to determine the appropriate duct dimensions for your specific project. Several software programs and industry standards assist in these calculations.

Accurate airflow measurement is critical during commissioning and routine maintenance. Using tools like anemometers and pressure gauges, technicians can verify that the designed airflow rates are achieved in each zone. Discrepancies, such as insufficient airflow to a particular area, can point to issues like restricted airflow due to accumulated dust or debris within the ductwork, or perhaps a malfunctioning damper. Addressing these issues promptly is crucial for maintaining optimal system performance. Proactive maintenance, including regular inspections and cleaning, minimizes these issues and ensures the long-term health and efficiency of the system.

Pressure within a duct system influences the system's efficiency and its ability to achieve desired performance metrics. High pressure in the ductwork can lead to increased energy consumption by the fan, while too low a pressure may result in insufficient airflow delivery to the spaces being conditioned. Finding the optimal balance requires the use of proper calculations and understanding of pressure-drop relationships throughout the system. This necessitates a thorough understanding of friction losses, which impede airflow due to resistance within the ductwork itself.

The selection of appropriate duct materials also significantly impacts airflow and pressure. Materials such as galvanized steel, flexible duct, and fiberglass offer different levels of resistance to airflow. Choosing the right material balances cost considerations with the implications for system performance. Furthermore, the geometry of the ductwork itself plays a significant role; sharp bends and sudden changes in cross-sectional area increase friction losses, which should be addressed in the design stage. A well-planned duct design minimizes these losses, ensuring effective and efficient air distribution throughout the building. Detailed specifications of duct sizes and materials form a vital part of your project's complete documentation and drawings.

Proper system balancing ensures that all zones receive the designed airflow. This often involves adjusting dampers within the ductwork, carefully controlling the volume of air directed to each area. This is especially important in buildings with variable occupancy patterns, where airflow requirements can change during the day. System balancing requires specialized tools and expertise. It may also require adjustments to the AHU's fan speed to compensate for pressure imbalances in the ductwork. The result of correctly balancing your system is efficient and even temperature distribution, which enhances energy conservation and improves occupant comfort.

Troubleshooting airflow and pressure problems requires a systematic approach. Begin by reviewing the system's design documentation to understand the intended airflow rates and pressure relationships. Then, use measuring instruments to verify the actual airflow and pressure values. Any discrepancies should be investigated further, which often involves checking for obstructions within the ductwork, examining the operation of dampers, and verifying the proper function of the AHU. Remember, thorough documentation of all measurements and observations is crucial for effective troubleshooting and problem resolution. This rigorous process minimizes potential downtime and ensures system efficiency.

3.3 Ventilation Strategies

Effective ventilation is crucial for maintaining healthy indoor air quality and thermal comfort. It's not simply about moving air; it's about strategically managing airflow to dilute pollutants, remove moisture, and control temperature and humidity levels. Understanding the different ventilation strategies allows for the creation of systems that are both effective and energy efficient, aligning perfectly with modern sustainable building design principles.

Proper ventilation involves more than just opening windows. For larger buildings, sophisticated systems are needed to control air exchange effectively. These systems range from simple exhaust fans removing stale air to complex networks of supply and return ducts, employing various ventilation methods. Each method, from natural ventilation to mechanical systems, offers specific advantages and limitations that must be considered within the context of the building and its occupants' needs. Choosing the right ventilation strategy depends on factors such as building size, occupancy density, climate, and budget considerations.

Natural ventilation relies on natural forces like wind and stack effect to drive airflow. It's often the most energy-efficient option, particularly in climates with consistent breezes and temperature differences between inside and outside air. However, it lacks control and predictability, making it less suitable for climates with unpredictable weather or for buildings with stringent air quality requirements. Effective natural ventilation often involves strategic placement of openings, such as windows and vents, designed to maximize airflow while minimizing unwanted drafts or heat transfer. Careful consideration of prevailing wind direction and solar orientation is essential to maximize its efficiency.

Mechanical ventilation, on the other hand, offers precise control over airflow rates and air quality. It uses fans to supply or exhaust air, enabling superior indoor air quality management regardless of outdoor conditions. There are several subtypes within mechanical ventilation: exhaust ventilation removes stale or contaminated air, supply ventilation introduces fresh air, and balanced ventilation combines both supply and exhaust to achieve a controlled air exchange. The choice between these depends on the specific needs of the building and the desired level of control. Mechanical ventilation systems are commonly integrated with air filtration systems to remove pollutants and allergens, further enhancing indoor air quality.

One critical aspect of ventilation is determining the necessary airflow rates. These rates are typically calculated using various standards and guidelines, often depending on the type of occupancy and the potential for pollutant generation. Occupancy levels are a crucial factor, dictating the amount of fresh air needed to maintain acceptable air quality. This is particularly crucial in densely populated spaces like classrooms, offices, or manufacturing facilities. Accurate calculations require considering factors such as the number of occupants, the area of the space, and the anticipated generation rate of pollutants, including carbon dioxide, volatile organic compounds (VOCs), and other contaminants. Using appropriate ventilation standards ensures the system will consistently meet the requirements of the building's use.

Designing an effective ventilation system requires careful consideration of various factors, including ductwork design, fan selection, and control strategies. The correct sizing of ducts to minimize pressure loss and ensure even airflow throughout the system is crucial. Incorrectly sized ducts could lead to uneven distribution, resulting in stagnant air in certain zones and potential discomfort or poor air quality. Appropriate fan selection is paramount for efficient and effective air movement. Selecting the right type and size of fan ensures proper air pressure throughout the duct system. It also balances air exchange against energy consumption, providing the balance between comfortable air and responsible resource use. Finally, implementing effective control strategies, such as using sensors to monitor air quality and adjust airflow rates automatically, ensures optimal performance and energy efficiency.

Integrating ventilation systems with other building systems, such as heating and cooling, is essential for overall energy efficiency. This often involves designing the system to leverage the thermal properties of the incoming air. A well-integrated system can reduce energy consumption associated with heating or cooling, promoting energy savings. This involves considerations for preheating or precooling incoming air in order to reduce the load on the heating and cooling systems. For instance, using heat recovery ventilators (HRVs) can transfer heat from the exhaust air to the incoming fresh air, significantly improving energy efficiency. Effective integration requires careful coordination between the ventilation system design and the design of other building systems.

Moreover, regular maintenance is essential for long-term performance and energy efficiency. Filters require periodic replacement to maintain air quality and prevent the build-up of pollutants. Regular inspection of the entire system helps catch potential problems early, minimizing downtime and ensuring the system is running at peak efficiency. A well-maintained ventilation system will continue to provide its designed benefits. Without proper maintenance, the overall efficacy of the ventilation strategy diminishes over time, rendering the initial design less efficient and reliable. Establishing a comprehensive maintenance schedule is crucial for the long-term health and efficiency of the system. The lifespan of HVAC equipment is directly affected by proper maintenance, leading to long-term cost savings and reducing the environmental impact of premature equipment replacement.

3.4 System Balancing

Balancing an HVAC system ensures consistent airflow throughout a building, crucial for maintaining comfortable temperatures and proper ventilation. Poorly balanced systems lead to uneven heating or cooling, wasted energy, and potential equipment damage. Think of it like a complex plumbing system; if the pressure isn't right in each branch, some areas will suffer while others are overwhelmed. Achieving balance requires careful attention to detail and methodical measurements.

The process typically begins with a thorough site survey, identifying all terminal units, such as diffusers, registers, and grilles. This involves noting their locations, sizes, and types. Accurate measurements are paramount, using tools like anemometers or pressure gauges to determine the actual airflow at each point. You need to understand the system's design, consulting blueprints and specifications to grasp the intended airflow patterns. Discrepancies between the design

and the actual measurements often signal the need for adjustments. These initial steps lay the foundation for a successful balancing effort, highlighting the importance of preparation.
Next, you'll use this data to calculate the airflow at each terminal unit. This might involve sophisticated calculations, or it may simply require adjusting dampers to achieve the design values. Remember, dampers are the primary tool in this process; they regulate airflow by partially obstructing or opening the ducts. Each adjustment requires careful consideration, as changing one damper's position can influence airflow elsewhere in the system. This iterative process demands patience and precision, often requiring multiple adjustments and remeasurements to achieve the desired balance. It's not a 'set-it-and-forget-it' process; it's a carefully orchestrated dance of adjustments.
Balancing involves more than just airflows; it also necessitates scrutinizing pressure differences within the system. Excessive pressure drops indicate problems like leaks, blockages, or poorly designed ductwork. Identifying and resolving these issues is vital for both balance and energy efficiency. A system operating under high pressure consumes more energy to overcome the resistance, adding unnecessary costs and reducing the system's overall performance. Finding these pres-

sure imbalances can sometimes lead to finding other problems with the system, further highlighting the importance of thoroughness.

Once you've achieved satisfactory airflow and pressure distribution, the final step involves documenting the results. This usually involves creating a detailed report indicating the measured airflow and pressure at each terminal unit, along with any adjustments made. This documentation serves as a vital record for future reference, aiding in troubleshooting or maintenance. It's a valuable resource for anyone working on the system later, as it reveals the current system balance and all the necessary design data. It also provides historical information, valuable for long-term maintenance and planning.

Beyond basic balancing, advanced techniques and technologies enhance accuracy and efficiency. Computerized balancing systems, for instance, automate many of the measurement and adjustment processes. These systems can significantly speed up the process, allowing for rapid and precise adjustments. Yet, even with advanced tools, a solid understanding of the underlying principles remains essential, as technology only enhances, not replaces, professional judgment. A skilled technician will always be needed to interpret the data and make informed decisions.

Remember, system balancing isn't a one-time event. As buildings age, or as equipment is replaced or modified, the system's balance can shift. Regular re-balancing ensures continued efficiency and performance. This is particularly true in buildings experiencing significant changes in occupancy or use patterns. Regular inspections and maintenance are crucial for long-term performance, ensuring the system continues functioning at its best. This aspect of ongoing maintenance underlines the fact that balancing is an essential part of long-term HVAC management. Proactive maintenance ensures longevity and cost savings.

Finally, always prioritize safety. When working with HVAC systems, especially those involving high pressures or moving parts, adhere strictly to safety protocols. Proper personal protective equipment and awareness of potential hazards are crucial. Never attempt adjustments without adequate training and experience; incorrect procedures can lead to serious accidents. Prioritizing safety ensures a safe and productive working environment. Remember, experience, training and attention to detail are more critical to effective and safe system balancing than any single tool.

3.5 Troubleshooting

Low airflow is a common issue. Check the air filter; a clogged filter drastically restricts airflow, leading to reduced cooling or heating capacity and potentially damaging the system's components. Regular filter changes, following the manufacturer's recommendations, are essential preventative maintenance. Inspect the ductwork for leaks or blockages; even small gaps can significantly reduce airflow efficiency. Seal any leaks with appropriate sealant, and remove any obstructions. Proper duct sizing during the initial design phase is crucial to prevent these problems, but addressing them after the fact is often necessary for optimal system performance. If the system still isn't performing optimally, you may need professional help with ductwork cleaning or repair. Remember to always prioritize safety when inspecting or working with ductwork, as it may contain sharp edges or insulation materials that require careful handling.

Inadequate heating or cooling can stem from several sources. Verify that the thermostat is set correctly and functioning properly. A malfunctioning thermostat could be sending incorrect signals to the system, preventing it from operating at its full potential. Consider replacing it if necessary. Check the system's refrigerant levels; insufficient refrigerant drastically impacts cooling capacity. Refrigerant leaks require professional attention and should be addressed promptly to prevent further damage. If the issue persists despite these checks, you might have a faulty heating element, compressor, or other internal component, requiring a professional HVAC technician's expertise for diagnosis and repair. A detailed inspection will help identify the specific cause of the problem and the necessary actions for a swift resolution.

Uneven temperatures across zones suggest an imbalance in the air distribution system. This problem might be due to incorrectly sized or improperly installed ducts. Consider that a poorly designed air distribution network, or one lacking regular maintenance, can cause some areas to be significantly hotter or colder than others. A thorough assessment of the air distribution system, potentially including a professional balancing, often resolves this issue. Ensure that all vents and registers are properly open and unobstructed. Closed or partially closed vents can restrict airflow to certain zones, resulting in uneven temperatures. The solution requires a systematic check of all vents throughout the system. This systematic approach helps ensure no area is missed.

Strange noises emanating from the HVAC system usually indicate a problem that should be addressed promptly. A rattling sound could result from loose components or debris within the system. A high-pitched squeal often points towards a problem with the blower motor or bearings. A grinding sound may signify problems within the compressor or other mechanical components. These sounds are often indicators of more serious underlying problems, and professional help should be sought. Delaying repairs might cause further damage and potentially more expensive fixes. Don't attempt DIY repairs unless you have the skills and experience to do so safely.

Excessive energy consumption is a clear sign of inefficiency. A leaking refrigerant charge, incorrect airflow, or a malfunctioning thermostat can all contribute to significantly higher energy bills. Regular maintenance, including filter changes, and thorough inspections help prevent this issue. Consider upgrading to a more energy-efficient system, especially if your current system is outdated. These improvements not only save energy but also often lead to more effective heating and cooling, improving overall comfort. Implementing energy-saving practices, such as proper insulation, can also reduce the system's energy consumption. Professional advice regarding system upgrades and energy conservation measures can be highly beneficial, especially for older systems. Remember that investing in energy efficiency not only improves the system's performance, but also reduces operational expenses.

Finally, water leaks around the system are a serious issue, potentially causing water damage and other problems. Check for condensation leaks from the condensate drain line. Clogs in the drain line, often due to dirt or debris accumulation, can easily lead to water buildup and overflow. Clean the drain line regularly to prevent this common problem. If the leak persists, it might indicate a more significant issue that requires professional attention, potentially involving a refrigerant leak or a malfunctioning component. Immediate attention is crucial to prevent serious damage to the building structure and any potential health hazards associated with leaks. A swift resolution to these leaks prevents further issues and potentially substantial repair costs. Professional assessment is strongly recommended if you are not confident in identifying the source and performing the necessary repairs yourself.

4 Refrigeration Cycles

The vapor-compression refrigeration cycle forms the backbone of most cooling systems. Understanding its components – the compressor, condenser, expansion valve, and evaporator – is crucial for any HVAC professional. The compressor, a powerhouse of the system, increases the refrigerant's pressure and temperature, converting it from a low-pressure, low-temperature vapor into a high-pressure, high-temperature vapor. This pressurized refrigerant then flows to the condenser, where it releases heat to the surroundings, transitioning from a hot, high-pressure gas to a high-pressure liquid. This heat rejection is vital, ensuring the system effectively removes heat from the space being cooled. The now high-pressure liquid refrigerant subsequently moves through the expansion valve, a critical component responsible for a significant pressure drop. This pressure reduction leads to a corresponding temperature drop, converting the high-pressure liquid into a low-pressure, low-temperature mixture of liquid and vapor. Finally, this low-pressure, low-temperature mixture enters the evaporator, where it absorbs heat from the space to be cooled, completing the cycle by evaporating into a low-pressure vapor. This continuous cycle ensures that heat is constantly removed from the desired area, maintaining a comfortable and controlled temperature. Efficiency in each stage significantly impacts the overall system performance.

Refrigerant selection plays a vital role in cycle efficiency and environmental impact. Numerous refrigerants exist, each with specific properties, advantages, and disadvantages. Factors such as their thermodynamic properties, environmental impact (Ozone Depletion Potential – ODP and Global Warming Potential – GWP), flammability, and toxicity must be carefully considered. Modern regulations and best practices strongly emphasize environmentally friendly refrigerants with low GWP values, minimizing the impact on the planet. For example, hydrofluoroolefins (HFOs) have emerged as a promising alternative to traditional refrigerants, offering a balance between performance and environmental responsibility. Proper refrigerant selection requires a comprehensive understanding of their characteristics and the specific application requirements. It's not simply a matter of choosing the most efficient refrigerant; a holistic assessment considering environmental impact and safety standards is absolutely necessary. Improper refrigerant selection can lead to reduced efficiency, increased operational costs, and potential environmental damage.

The efficiency of a vapor-compression cycle is paramount. Several factors significantly influence this efficiency, including the compressor's efficiency, the effectiveness of heat exchange in the condenser and evaporator, and the pressure drop across the expansion valve. Optimizing these components and the overall system design is crucial for maximizing performance and minimizing energy consumption. A well-designed system minimizes pressure drops, ensuring efficient refrigerant flow and reducing energy losses. Similarly, effective heat exchangers (condenser and evaporator) transfer heat efficiently, maximizing the amount of heat removed from the conditioned space and released to the surroundings. Advanced technologies, such as variable-speed compressors and electronically-expanded valves, further enhance efficiency by allowing for precise control and adaptation to varying load demands. Regular maintenance and proper system operation are also key contributors to sustained high efficiency. Neglecting these aspects can lead to decreased performance and increased energy costs over the system's lifetime.

Beyond the standard vapor-compression cycle, other refrigeration cycles exist, each with its own application and characteristics. Absorption refrigeration, for instance, utilizes heat instead of electricity to drive the refrigeration process. This technology is well-suited for applications where heat is readily available, such as waste heat recovery systems or solar-powered cooling. Another alternative is the thermoelectric refrigeration cycle, which uses the Peltier effect to achieve cooling. While relatively less efficient than vapor-compression systems, thermoelectric cooling offers advantages in its compactness and lack of moving parts. Selecting the appropriate refrigeration cycle depends heavily on the specific application, considering factors such as energy source availability, desired cooling capacity, and environmental impact. Each cycle comes with its own set of advantages and limitations, and choosing correctly is essential for optimal performance and energy efficiency. Understanding these different methods is necessary for professionals to offer their clients the most efficient and sustainable solutions.

Troubleshooting refrigeration cycles necessitates a systematic approach. Analyzing system pressures, temperatures, and refrigerant flow rates provides crucial insights into potential issues. Leaks are a common problem, requiring careful detection and repair to maintain system integrity and efficiency. Compressor malfunctions can significantly impact performance, necessitating diagnosis and potential replacement. Problems with the expansion valve or heat exchangers can also lead to reduced efficiency or complete system failure. A thorough understanding of the cycle's operating principles and component functionalities is vital for effective troubleshooting and timely resolution of any issues. This systematic approach not only ensures the system's proper functioning but also prolongs its lifespan and minimizes operational costs. Proper diagnostics and timely repairs prevent escalating problems.

4.1 Vapor Compression Cycles

The vapor-compression refrigeration cycle is the cornerstone of most modern air conditioning and refrigeration systems. Understanding its intricacies is fundamental for any HVAC professional. This cycle, a closed loop, relies on a refrigerant's unique properties to move heat from a lower-temperature space to a higher-temperature one. Imagine it like a tireless worker, constantly transferring thermal energy against a natural gradient.

This seemingly simple process involves four key components: the compressor, condenser, expansion valve, and evaporator. The compressor, a powerful pump, increases the refrigerant's pressure and temperature significantly. It's the heart of the system, responsible for driving the entire cycle. Think of it as the engine powering the heat transfer. This high-pressure, high-temperature refrigerant then flows into the condenser.

Within the condenser, heat is transferred from the high-pressure refrigerant to the surrounding environment. This typically involves a network of fins that maximize surface area for efficient heat exchange, essentially allowing the refrigerant to cool and condense into a liquid state. The air moving across the condenser coils carries away this released heat. Efficient condenser design is crucial for effective cooling and system performance. Consider the size and placement of the condenser crucial for its efficient operation.

After condensing, the high-pressure liquid refrigerant passes through the expansion valve. This component causes a dramatic pressure drop, which simultaneously lowers the refrigerant's temperature. The sudden pressure reduction converts the liquid refrigerant into a mixture of liquid and vapor, a critical step for effective heat absorption in the subsequent stage.

Think of this as strategically lowering the refrigerant's potential energy.

Finally, the low-pressure, low-temperature refrigerant enters the evaporator. Here, the refrigerant absorbs heat from the space being cooled. This heat absorption causes the refrigerant to evaporate, completing the cycle. The resulting low-pressure vapor then returns to the compressor, initiating another cycle. The evaporator's surface area and design are paramount to its effective heat-absorbing capacity. The size, positioning, and airflow are important aspects to consider.

Refrigerant selection is a critical aspect of vapor-compression cycle design. Different refrigerants possess varying thermodynamic properties and environmental impacts. Some refrigerants, while efficient, contribute to ozone depletion or global warming. Therefore, the choice of refrigerant often involves balancing performance with environmental responsibility. Regulations regarding refrigerants are constantly evolving, making it vital to stay updated.

Several factors influence the efficiency of a vapor-compression cycle. The compressor's performance, for instance, directly affects energy consumption. Heat transfer effectiveness in the condenser and evaporator also significantly impacts the overall efficiency. Proper design and maintenance are critical to maximizing the system's efficiency. The system's efficiency directly impacts the system's overall cost-effectiveness.

Understanding the pressure-enthalpy diagram (P-h diagram) is essential for analyzing and optimizing vapor-compression cycles. This diagram graphically represents the thermodynamic properties of the refrigerant throughout the cycle. It allows engineers to visualize the changes in pressure, temperature, and enthalpy at each stage, enabling performance assessment and optimization. Accurate interpretation of this diagram is an invaluable skill.

Analyzing the performance of a vapor-compression cycle involves calculating various parameters like the coefficient of performance (COP) which is a key metric. A higher COP indicates better energy efficiency. Various losses and inefficiencies within the cycle, such as pressure drops and heat losses, need to be considered when evaluating performance. Optimizing these aspects through design and maintenance is essential for cost-effectiveness.

Troubleshooting problems in vapor-compression systems often involves a systematic approach, beginning with observing the system's behavior and analyzing pressure and temperature readings. Identifying component failures or inefficiencies, such as refrigerant leaks or malfunctioning components, are usually approached with specific procedures. Regular system maintenance is paramount in avoiding major breakdowns and ensuring long-term reliability. Proactive maintenance is crucial for maintaining the efficiency and longevity of the system.

The proper sizing of a vapor-compression system is also critical for optimum performance. This requires careful consideration of the load (the amount of heat to be removed) and the ambient conditions. Undersized systems struggle to meet the cooling demand, leading to reduced efficiency and poor performance; oversized systems operate inefficiently due to short cycling. A proper design addresses these parameters to ensure long-term cost-effectiveness.

Finally, advancements in vapor-compression technology continue to improve efficiency and reduce environmental impact. New refrigerants with lower global warming potential are being developed, and improvements in compressor and heat exchanger design aim at optimizing energy consumption. Keeping abreast of these advancements is vital for any professional engaged in this field. Staying updated allows professionals to optimize design and maintenance for best outcomes.

4.2 Refrigerant Properties

Understanding a refrigerant's properties is fundamental to designing efficient and safe HVAC systems. These properties dictate the refrigerant's behavior within the refrigeration cycle, impacting system performance, energy consumption, and environmental impact. Selecting the right refrigerant is crucial, a process demanding careful consideration of various factors.

Refrigerants are classified by their chemical composition and thermodynamic properties. Some common types include chlorofluorocarbons (CFCs), hydrochlorofluorocarbons (HCFCs), hydrofluorocarbons (HFCs), and natural refrigerants like ammonia and carbon dioxide. Each type exhibits distinct advantages and disadvantages concerning performance, safety, and environmental impact. For example, while CFCs and HCFCs offer excellent thermodynamic properties, their ozone depletion potential led to their phase-out under the Montreal Protocol.

The key thermodynamic properties to consider include saturation temperature, pressure, enthalpy, and specific volume. The saturation temperature-pressure relationship defines the refrigerant's boiling and condensation points at different pressures, critical for determining the operating pressures within the refrigeration cycle. Understanding this relationship is essential for compressor selection and system sizing. High-pressure refrigerants may demand more robust components.

Enthalpy, representing the total heat content, dictates the heat transfer during evaporation and condensation. Changes in enthalpy reflect the amount of heat absorbed or released, essential for calculating the refrigeration capacity and energy efficiency of the system. Analyzing enthalpy-entropy diagrams helps visualize these changes, crucial for accurate thermodynamic calculations.

Specific volume, representing the volume occupied per unit mass, affects the refrigerant flow rate and compressor size. Refrigerants with lower specific volumes require smaller compressors for the same mass flow rate, potentially leading to cost savings and increased efficiency. However, this must be balanced against other properties.

Beyond thermodynamic properties, safety aspects are paramount. Toxicity and flammability are critical considerations. Ammonia, for instance, is highly toxic, requiring stringent safety measures during handling and installation. Conversely, some HFCs may be less toxic but still flammable, demanding careful system design. These safety considerations significantly influence refrigerant selection.

Environmental impact is another crucial factor. Ozone depletion potential (ODP) and global warming potential (GWP) are metrics assessing a refrigerant's contribution to environmental problems. The ODP measures a refrigerant's ability to deplete the ozone layer, while the GWP reflects its contribution to climate change. Modern regulations increasingly favor refrigerants with low ODP and GWP values. For instance, many modern designs use HFO refrigerants, designed to have minimal environmental effects.

The selection process involves balancing these properties. A refrigerant with excellent thermodynamic properties might have a high GWP, requiring a trade-off. For example, using R-134a, a hydrofluorocarbon, is a popular choice for many systems, but it features a GWP that causes environmental concerns prompting consideration of other refrigerants. Therefore, thorough lifecycle assessments, accounting for energy efficiency and environmental impact throughout the refrigerant's lifespan, are critical.

Choosing the appropriate refrigerant requires careful consideration and a deep understanding of the system's requirements, such as operating temperature and pressure, capacity, and environmental regulations. Ignoring the properties of different refrigerants can result in inefficient, unsafe, or environmentally damaging systems. Therefore, utilizing reputable databases and design software, that contain up-to-date information on refrigerant properties, and adhering to industry best practices are strongly recommended. Using updated and reliable information is a key to successful HVAC system design. Remember that the choice of refrigerant directly impacts several aspects of the design and operation of an HVAC system and should be chosen with care.

Finally, remember that refrigerant properties are complex and interconnected. Accurate design requires the use of property data from reputable sources and sophisticated design software. This ensures that the refrigerant choice aligns with operational requirements and environmental sustainability, resulting in an efficient and responsible system. This detailed consideration of refrigerant properties is essential to achieve optimal HVAC system performance.

5 HVAC Controls and Automation

Precise temperature and humidity control is paramount in modern buildings, impacting occupant comfort, product quality (in industrial settings), and overall energy efficiency. Achieving this requires a sophisticated understanding and implementation of HVAC controls and automation. This goes beyond simple thermostats; we're talking about integrated systems that monitor numerous parameters and react intelligently to changing conditions.

The core of any automated HVAC system lies in its sensors and actuators. Sensors are the eyes and ears, constantly monitoring conditions like temperature, humidity, airflow, and pressure at various points throughout the system. These data points provide real-time feedback that informs the system's actions. Actuators, on the other hand, are the muscles, responsible for executing changes based on the sensor readings. They control dampers, valves, and the speed of fans and compressors, adjusting the HVAC system's output to maintain optimal conditions. Consider, for example, a system using multiple temperature sensors across a large warehouse; if one zone is too cold, only the heating in that particular zone will adjust, avoiding unnecessary energy expenditure in other areas that remain at a comfortable temperature.

Control strategies form the brains of the operation. Proportional-Integral-Derivative (PID) control is a widely used algorithm that provides a flexible approach to maintaining set points. PID controllers continuously compare the measured value with the desired value (the setpoint) and adjust the actuator's output accordingly. The proportional component reacts to the current error, the integral component addresses accumulated error over time, and the derivative component anticipates future error based on the rate of change. More advanced control systems may use predictive algorithms which forecast future energy needs based on occupancy schedules and weather forecasts. This proactive approach can significantly reduce energy consumption while maintaining comfort levels.

Building Management Systems (BMS) represent a significant

advancement in HVAC control. These centralized systems integrate data from various sensors and controllers across the entire building, providing a holistic view of the HVAC system's performance. BMS not only monitor conditions in real-time, allowing for prompt identification of problems, but also provide historical data for performance analysis and optimization. They often include features like energy dashboards displaying energy consumption patterns, allowing facility managers to identify areas where improvements can be made. Furthermore, modern BMS frequently incorporate features that allow for remote monitoring and control, facilitating proactive maintenance and swift response to unexpected events.

Effective use of BMS can lead to substantial energy savings. By optimizing the performance of individual HVAC units and coordinating their operation, a BMS can reduce energy waste associated with over-heating or over-cooling. Furthermore, some BMS platforms allow for the integration of renewable energy sources, further enhancing the building's sustainability profile. The capability to monitor equipment performance and predict potential failures provides the opportunity for timely maintenance, thus preventing major breakdowns and associated disruptions. The ability to schedule operations, for example, reduces energy usage outside of normal working hours.

Beyond typical temperature and humidity control, sophisticated systems are beginning to integrate other factors, creating a more dynamic and holistic approach to indoor environmental quality. We are seeing the increasing integration of CO2 sensors, monitoring occupant density and providing ventilation adjustments based on occupancy levels. Similarly, air quality sensors can detect volatile organic compounds (VOCs) and other pollutants, allowing for proactive ventilation strategies to mitigate their effects. This approach moves beyond just basic temperature regulation and encompasses a much broader concept of optimizing occupant well-being. This reflects a trend toward more human-centric designs, prioritizing the health and productivity of building occupants.

Integrating these advanced control systems into existing buildings can be a significant undertaking, requiring careful planning and coordination. Retrofitting older systems may involve upgrading existing components or replacing them entirely. This calls for a detailed assessment of the current system, identification of areas for improvement, and selection of appropriate technologies. Prioritizing energy efficiency during the upgrade process is critical, ensuring long-term cost savings and a smaller environmental impact. It's also important to ensure the BMS system is fully integrated with other building systems, allowing for streamlined management and coordinated control.

The future of HVAC controls lies in further integration of data, advanced analytics, and artificial intelligence. Machine learning techniques can be used to optimize control algorithms based on historical data, leading to even more efficient and responsive systems. The predictive capabilities of these systems will significantly improve proactive maintenance and prevent costly breakdowns. The shift towards smart buildings promises to optimize not only the comfort and productivity of building occupants but also the overall sustainability of building operations, improving resource management and minimizing environmental impacts. A holistic approach, considering all aspects of building operations, is essential for realizing the full potential of these advanced technologies.

5.1 Sensors and Actuators

Sensors are the eyes and ears of an HVAC system, constantly monitoring conditions to ensure optimal performance. They measure temperature, humidity, pressure, airflow, and other critical parameters. Think of a thermostat; it's a simple yet essential sensor, providing feedback to the system's controller. More sophisticated systems use a wide array of sensors, from thermocouples for precise temperature measurement to pressure transducers for monitoring airflow in ducts. The data these sensors collect is crucial for making informed decisions about system operation, identifying potential problems before they escalate, and optimizing energy efficiency. Understanding the capabilities and limitations of various sensor types is critical for selecting the right ones for any given application. A poorly chosen sensor can lead to inaccurate readings, improper system operation, and ultimately, discomfort and increased energy costs. For instance, a low-resolution temperature sensor may not accurately reflect subtle changes in the environment, leading to inefficient heating or cooling. Conversely, a sensor with excessive sensitivity might trigger unnecessary adjustments, increasing energy consumption without noticeable improvement in comfort. Careful sensor selection is a key aspect of system design, ensuring accuracy, reliability and cost-effectiveness. The choice depends on several factors: the required accuracy, the range of measurement, the environmental conditions, the cost, and the ease of installation and maintenance. The right sensor is an investment, paying for itself over time through better performance and reduced energy bills.

Actuators are the muscles of the HVAC system, responding to the signals received from sensors to make the necessary adjustments. They are the components that actually control the system's operation. A simple example is a valve that controls the flow of hot or cold water into a radiator or the opening and closing of dampers in air ducts. More advanced systems utilize electric motors for controlling fan speed, compressors, and pumps. Actuators are not just on/off switches; many are capable of modulating output, allowing for fine-grained control of system variables. This modulation is crucial for efficient operation, preventing excessive energy use and improving occupant comfort. For instance, a modulating valve can adjust water flow precisely based on the temperature difference between the supply and return lines, ensuring optimal heating or cooling. Without precise actuator control, the system might overshoot or undershoot its setpoints, leading to energy waste and discomfort. The choice of actuators is as important as sensor selection. Actuators need to be powerful enough to handle their loads, while simultaneously being reliable and efficient. Consider factors like response time, power consumption, and durability when selecting actuators. Choosing the right actuator ensures the system is reliable, efficient, and able to handle the demands of its environment. Just as a weak muscle wouldn't be suitable for a demanding task, an undersized actuator may struggle to perform its role, leading to performance issues.

The integration of sensors and actuators is crucial for creating a closed-loop control system. This system constantly monitors the environment, compares the measured values to the desired setpoints, and makes adjustments as needed. This closed-loop system uses feedback mechanisms to adjust operations continuously, creating a dynamic response to environmental changes. The system's performance is directly related to the quality of the sensors and actuators. A precise and reliable sensor coupled with a robust and responsive actuator ensures efficient control, maintaining optimal comfort while minimizing energy waste. The use of feedback allows the system to be self-correcting, adapting dynamically to changing conditions, which ensures optimum performance. This is especially important in buildings with fluctuating occupancy patterns or external conditions such as solar radiation and outdoor temperature. A well-integrated system can anticipate changes and adjust proactively, minimizing deviations from the desired comfort levels. In complex HVAC systems, the integration of multiple sensors and actuators requires sophisticated control algorithms, often involving programmable logic controllers (PLCs) or building management systems (BMS). These systems allow for centralized monitoring and control of multiple zones, ensuring coordinated operation and optimal energy efficiency. This is an essential consideration as the system complexity grows, ensuring a streamlined approach to the overall control system.

Modern HVAC systems rely heavily on digital control technology, with sensors and actuators connected through a network to a central controller. The use of digital technologies also opens up the possibility of remote monitoring and control, enabling proactive maintenance and adjustments. Data logging capabilities allow for trend analysis, identifying potential issues before they escalate and facilitating improvements in system design and operation. This allows for greater control over the entire HVAC system and improves efficiency. This ability to monitor and adjust remotely also improves maintenance by enabling rapid response to issues and reducing downtime. This data collection capability transforms the system from a reactive to a proactive one, maximizing efficiency, and minimizing discomfort. The benefits are significant, from optimized energy consumption to increased comfort and reduced maintenance costs. The integration of smart sensors and advanced control algorithms is driving the development of increasingly intelligent and energy-efficient HVAC systems, enhancing both performance and sustainability. This constant monitoring and control contribute to long-term efficiency and sustainability.

5.2 Control Strategies

Direct digital control (DDC) systems offer precise control over HVAC components. These systems utilize microprocessors to monitor and adjust variables like temperature, humidity, and airflow based on programmed setpoints. The ability to finely tune these parameters provides significant energy savings and improved comfort. For instance, a DDC system can modulate a variable-speed fan based on real-time occupancy data, reducing energy consumption during periods of low occupancy. Furthermore, advanced DDC systems can incorporate predictive control algorithms, anticipating future needs and proactively adjusting system operation. This proactive approach minimizes energy waste and maintains optimal conditions.

PID controllers remain a cornerstone of HVAC control strategies. This classic three-term controller adjusts output based on proportional, integral, and derivative components of the error signal—the difference between the setpoint and the actual value. The proportional term provides immediate response to errors. The integral term eliminates persistent offsets, while the derivative term anticipates future errors. Proper tuning of the PID controller's gain parameters is crucial for optimal performance. Overly aggressive tuning can lead to instability and oscillations, while overly conservative tuning results in slow response times and potentially missed setpoints. Sophisticated tuning methods, such as Ziegler-Nichols, can assist in finding the optimal gains.

Sequencing strategies optimize the operation of multiple HVAC units within a larger system. This involves controlling the startup and shutdown of individual units based on load demand. For instance, a large building might employ multiple air handling units (AHUs). Instead of operating all AHUs continuously, a sequencing strategy would activate only the necessary units to meet the current cooling or heating load. This approach reduces energy consumption by avoiding the simultaneous operation of multiple units when only a portion of their capacity is required. Effective sequencing relies on accurate load forecasting and a robust control system that can manage multiple units efficiently.

Zone control enhances comfort and efficiency by individually controlling smaller areas within a building. This is achieved by dividing the building into distinct zones, each with its own thermostat and control system. This allows for precise temperature regulation in each zone, adapting to the specific needs of occupants in different areas. For example, a large office building may have separate zones for different floors or even individual offices. This zonal approach minimizes energy waste by avoiding over-cooling or over-heating unoccupied spaces. The implementation of zone control requires careful consideration of air distribution and the integration of multiple control systems.

Many modern HVAC systems incorporate sophisticated predictive control strategies. These algorithms use historical data, weather forecasts, and occupancy schedules to anticipate future heating and cooling demands. By predicting the required system output in advance, these systems can proactively adjust operation, minimizing energy consumption and improving comfort. For example, a predictive control system could pre-cool a building during off-peak hours in anticipation of high occupancy during the day. The effectiveness of predictive control heavily depends on the accuracy of the prediction models and the availability of relevant data. The initial investment in data collection and system integration might be substantial, but the long-term energy savings can be significant.

Fault detection and diagnostics (FDD) are critical aspects of modern HVAC control strategies. These systems use advanced algorithms to monitor system performance and detect potential problems before they escalate into major failures. FDD systems can identify anomalies in sensor readings, operational patterns, and component behavior, signaling potential issues to maintenance personnel. This proactive approach minimizes downtime, reduces repair costs, and improves overall system reliability. Sophisticated FDD systems utilize machine learning techniques to learn and adapt to the unique characteristics of each HVAC system. Regular maintenance and calibration are essential to ensure the accuracy and effectiveness of FDD systems. Their early detection of issues saves money and reduces maintenance problems.

Finally, building management systems (BMS) provide centralized control and monitoring of entire buildings' HVAC systems. These systems integrate data from various sensors and control devices, providing a comprehensive overview of building operations. BMS allows for remote monitoring, automated fault detection, and centralized scheduling. This integration improves energy efficiency by optimizing the operation of all HVAC equipment. A properly configured BMS provides operators with valuable insights into building performance. Additionally, BMS are increasingly incorporating advanced analytics capabilities to identify further opportunities for energy saving and performance improvements. The capacity of these systems greatly increases energy efficiency and reduces maintenance time.

5.3 Building Management Systems

Building management systems (BMS) are the central nervous systems of modern buildings, seamlessly integrating and optimizing the performance of all building systems, including HVAC. A well-designed BMS provides real-time monitoring of various parameters, allowing for proactive adjustments and minimizing energy consumption. Its importance cannot be overstated in today's world of increasingly complex and energy-intensive structures. Imagine a skyscraper, its countless HVAC units humming away. Without a BMS, managing such a system would be akin to conducting an orchestra without a conductor—chaotic and inefficient. A BMS brings order to this complexity.

Think of it as a sophisticated digital dashboard, displaying crucial information like temperature, humidity, air quality, and equipment status from all corners of the building. This data is invaluable. For example, detecting a malfunctioning chiller before it leads to a complete system failure prevents costly downtime and potential environmental damage. The timely intervention enabled by continuous monitoring is a significant advantage. But the benefits extend beyond simply identifying problems.

The real power of a BMS lies in its ability to proactively manage and optimize building systems for peak efficiency. By analyzing historical data and real-time conditions, the system can automatically adjust parameters, such as ventilation rates, to maintain optimal indoor comfort while reducing energy usage. For instance, if a room is unoccupied during off-peak hours, the BMS can automatically lower the temperature and reduce airflow, thereby saving considerable energy. This predictive capability is crucial for achieving sustainable building operation. It's not just about reaction; it's about anticipation and intelligent management.

Proper implementation is crucial for the system's effectiveness. Careful planning and design are essential to ensure the BMS integrates seamlessly with existing and future building systems. The selection of appropriate hardware and software is paramount. It should be scalable to accommodate future growth and technological advancements, and easily adaptable to changing needs and regulations. Think of it as laying the foundation for a technologically advanced future, not just meeting present requirements. Neglecting this aspect often leads to costly retrofits and limited functionality later on.

Effective training is often overlooked but is incredibly important. Building operators need comprehensive training on how to use and interpret the data provided by the BMS. This ensures they can effectively monitor the system, respond to alerts, and make informed decisions. A well-trained team is the key to harnessing the full potential of the system. Knowledge empowers individuals to maximize energy efficiency and maintain a healthy building environment. Without proper training, a sophisticated system is essentially useless.

Furthermore, integrating the BMS with other building systems, such as lighting and security, creates a holistic approach to building management. This integrated approach can lead to further optimization opportunities, enhancing efficiency and user experience. For example, the BMS could automatically adjust lighting based on occupancy and daylight availability, optimizing energy usage while maintaining appropriate light levels. This synergistic effect between different building systems is where the true power of a modern BMS lies. It's more than the sum of its parts; it's a synergistic system working in perfect harmony.

Finally, ongoing maintenance and updates are indispensable. Regular system checks, software updates, and calibration of sensors ensure the BMS remains accurate and reliable over time. Neglecting these crucial tasks can lead to inaccuracies, reduced functionality, and potential system failures. A properly maintained BMS is an ongoing investment, providing long-term benefits for building owners and operators. This ensures its long-term effectiveness, prevents costly repairs and downtime, and guarantees the continued accuracy and performance of all the monitored systems. It is a commitment to a consistently efficient and reliable system.

6 Energy Efficiency in HVAC

Improving HVAC energy efficiency is crucial for both environmental sustainability and economic viability. We're talking significant reductions in operational costs, a lessened carbon footprint, and a more responsible use of resources. Consider the potential savings—that's money directly back in your client's pocket, a substantial return on investment when implemented correctly.

A thorough energy audit forms the cornerstone of any improvement strategy. Don't just glance over the system; delve deep into its operation. Identify areas with significant energy loss, pinpointing precisely where improvements can generate the greatest impact. Are there leaks in the ductwork? Is the insulation inadequate? These seemingly minor issues can accumulate into significant energy drains over time. Detailed data logging and thermal imaging can highlight previously unseen inefficiencies. Remember that precision is key; you're not just aiming for general improvements, but focused, targeted optimization.

Investing in high-efficiency equipment pays dividends in the long run. Modern HVAC systems boast significantly improved Energy Efficiency Ratios (EER) and Seasonal Energy Efficiency Ratios (SEER) compared to older models. Replacing outdated equipment should be a top priority, particularly in buildings with significant energy consumption. Consider the lifecycle cost analysis: the initial investment, coupled with future energy savings, will quickly showcase the financial wisdom of such upgrades. Don't just focus on the upfront costs, evaluate the long-term value proposition. Choose equipment that aligns perfectly with the specific building needs and climate conditions, optimizing performance and minimizing energy wastage.

Beyond equipment upgrades, numerous energy conservation measures (ECMs) exist to improve efficiency. Simple improvements like optimizing thermostat settings and implementing occupancy sensors can yield substantial savings with minimal capital investment. Proper maintenance of your equipment is also a critical yet often overlooked area. Regular cleaning, filter changes, and preventative maintenance ensure peak performance and prevent efficiency-sapping issues. Don't underestimate the power of seemingly minor adjustments – their cumulative impact can be impressive. Educate your clients about these simple measures to ensure long-term effectiveness.

Integrating renewable energy sources into your HVAC system is a growing trend, and a vital step toward sustainability. Solar thermal systems can pre-heat water for use in radiant heating, or provide heat for absorption chillers. Similarly, geothermal energy offers consistent temperatures for heating and cooling, greatly reducing reliance on fossil fuels. Integrating these renewables, of course, requires careful system design and integration with existing infrastructure. Analyze the suitability of different renewable options; the optimal choice depends on location-specific factors, such as the availability of sunlight or geothermal resources. These systems, however, often require a larger upfront investment; a thorough cost-benefit analysis is necessary to justify their implementation.

Building codes and energy standards should guide your design choices. Staying abreast of current regulations is crucial, not merely for compliance, but also to incorporate the latest best practices in energy efficiency. These codes often incorporate energy modeling requirements; these provide valuable insights into the expected energy performance of the proposed system. Understanding these building regulations is more than just compliance; it demonstrates professionalism and commitment to sustainable practices, enhancing your credibility.

Lifecycle cost analysis (LCCA) is integral to your decision-making process. LCCA considers the total cost of ownership throughout the system's lifespan, encompassing initial investment, operating costs, maintenance, and potential replacement costs. This comprehensive approach avoids short-sighted decisions driven solely by upfront costs. When you provide clients with a clear LCCA, you demonstrate a responsible and financially astute approach. By showing the long-term return on investment for energy-efficient choices, you establish trust and solidify your professional reputation.

Finally, remember that energy efficiency is a holistic approach, demanding thorough analysis and careful planning. Consider every component of the HVAC system—from the equipment itself, to the ductwork, insulation, controls, and building envelope—to ensure optimal efficiency. Continuous monitoring and fine-tuning are necessary for maintaining peak performance over time. A well-designed, energy-efficient

HVAC system represents a substantial investment in the future, both environmentally and economically. This approach goes beyond compliance; it's about creating sustainable, high-performing systems that benefit both the environment and the client's bottom line. The long-term benefits will far outweigh the initial investment, cementing your reputation as a forward-thinking professional. Never underestimate the value of an expertly designed, energy-efficient HVAC system.

6.1 Energy Audits

Energy audits are crucial for optimizing HVAC systems' performance and reducing energy consumption. They provide a detailed assessment of a building's energy use, specifically pinpointing inefficiencies within the heating, ventilation, and air conditioning systems. This detailed analysis helps identify areas for improvement, leading to significant cost savings and reduced environmental impact. Remember, a thorough energy audit isn't just about finding problems; it's about crafting effective solutions.

The process typically begins with a preliminary assessment of the building's characteristics. Factors considered include the building's size, age, construction materials, occupancy levels, and climate zone. This initial overview helps establish a baseline for energy usage and identifies potential areas of concern. We then move on to a detailed examination of the HVAC equipment itself. This involves inspecting all components—from the chillers and boilers to the air handlers and ductwork—to check for leaks, malfunctions, and inefficiencies. Don't underestimate the value of a precise examination; often, small issues can snowball into considerable energy waste.

Following the equipment inspection, we perform thorough testing. This might involve using infrared cameras to detect thermal leaks, pressure gauges to measure airflow, and specialized meters to assess the efficiency of individual components. Remember to document everything meticulously; this data forms the core of your energy audit report and serves as the foundation for future improvements. Accurate data collection is non-negotiable. It allows for precise calculations and ensures the recommendations are realistic and effective. These detailed measurements help quantify the extent of energy loss and pinpoint the areas needing the most attention.

After the data is gathered and analyzed, we can generate a comprehensive report outlining the findings. This report serves as a roadmap for improvement, detailing the identified inefficiencies and outlining the recommended solutions. The report often includes specific, actionable recommendations for upgrades or repairs, prioritizing cost-effectiveness and long-term energy savings. It also contains projected energy savings and return on investment for each suggested improvement. Such clarity prevents costly mistakes and guarantees the client understands the potential gains.

Next, we need to discuss implementation. The audit report's recommendations need practical application. Depending on the findings, this might involve simple fixes like sealing air leaks in ductwork or replacing outdated equipment. This phase necessitates a careful balance between immediate fixes and long-term strategies. We must consider the financial implications, operational disruption, and the overall impact on the building's occupants. Each decision must be weighed carefully to ensure smooth implementation.

Finally, post-implementation monitoring is vital to confirm the effectiveness of the implemented changes. We must track energy consumption after the improvements have been made to verify the savings predicted in the audit report. Monitoring allows us to fine-tune our approach and ensure the system is operating as efficiently as possible. This ongoing observation not only verifies our work but also provides valuable feedback for future audits. Continuous improvement is key to optimal energy management. Therefore, regular monitoring and analysis are vital, offering opportunities for further adjustments and optimization. Don't underestimate the long-term benefits of this commitment.

Remember that an energy audit isn't a one-time event; it's an ongoing process. Regular audits, even if less intensive than the initial one, will allow for the early detection of potential problems and prevent energy waste from accumulating over time. The knowledge gained through this repetitive process strengthens understanding of the building's energy consumption patterns and allows for proactive rather than reactive maintenance. This continuous improvement cycle guarantees cost savings and enhanced efficiency. Consistent auditing promotes both short-term and long-term energy efficiency. Effective energy audits combine comprehensive testing, detailed analysis, and well-defined recommendations. The process is not solely about identifying problems; it's about developing practical and cost-effective solutions to improve the HVAC system's energy efficiency. The knowledge gained helps building owners and managers make informed decisions and realize substantial cost savings. Therefore, prioritize a detailed and well-executed energy audit to optimize your HVAC system's performance and ensure long-term sustainability. The meticulous effort yields significant returns in energy efficiency and reduced operating costs. This will ultimately contribute to a more sustainable and environmentally friendly built environment.

6.2 Energy Conservation Measures

Improving the efficiency of HVAC systems is crucial for both environmental sustainability and economic viability. Consider implementing variable-speed drives (VSDs) on your equipment. These devices allow for precise control of motor speed, matching output to the actual demand, rather than running at a constant, potentially excessive speed. This significantly reduces energy consumption, especially during periods of low heating or cooling demand. Furthermore, the reduced wear and tear from smoother operation extends equipment lifespan, resulting in long-term cost savings that can offset the initial investment. The precise control provided by VSDs allows for a more responsive system, immediately adjusting to changing conditions and preventing energy waste. Careful integration and programming of these systems within the overall building management system is crucial to realizing these benefits. Thorough training for operators and maintenance personnel are also necessary to fully optimize the systems performance.

Many buildings suffer from inefficient air distribution. Leaky ductwork represents a significant loss of conditioned air, forcing the system to work harder and consume more energy to maintain the desired temperatures. Regular inspection and sealing of ductwork, including both supply and return ducts, is essential. This isn't just a matter of finding large holes; minor leaks, even seemingly insignificant ones, accumulate to create a considerable energy drain over time. Using specialized tools to identify and pinpoint air leaks will allow for efficient repairs, minimizing disruption and maximizing the energy saving impact. The implementation of proper insulation along the ductwork route is also very important. The use of high-R-value insulation will reduce heat transfer, keeping conditioned air from escaping. Remember, energy efficiency measures are cumulative. Each small improvement contributes to the larger goal of significant energy savings and minimized carbon footprint.

Another potent strategy for energy conservation involves optimizing the building's thermal envelope. Efficient insulation of walls, roofs, and floors drastically reduces heat transfer, significantly lessening the load on the HVAC system. Proper insulation involves choosing the correct type and thickness of insulation materials based on the climate and the building's construction. Consider selecting high-performance materials with low thermal conductivity to maximize effectiveness. In addition to insulation, air sealing is equally crucial. Air leakage through cracks and gaps around windows, doors, and other building penetrations can negate much of the benefits gained from insulation. Therefore, carefully sealing all such openings significantly reduces energy loss. This task often involves the use of caulking, weatherstripping, and other appropriate sealing materials and techniques. Investing in high-performance windows with low U-values also contributes to reducing heat transfer through glazing, a significant source of heat loss in many buildings.

The selection of appropriate HVAC equipment is vital for energy efficiency. High-efficiency units, certified by organizations like the Air Conditioning, Heating, and Refrigeration Institute (AHRI), offer significant energy savings over less efficient models. The use of SEER (Seasonal Energy Efficiency Ratio) and EER (Energy Efficiency Ratio) ratings helps in evaluating the efficiency of air conditioning equipment. Similarly, AFUE (Annual Fuel Utilization Efficiency) is used to determine efficiency of furnaces. The higher the rating, the greater the energy savings. Consider units with advanced features like modulating gas valves, variable refrigerant flow (VRF) systems, or heat recovery ventilators to further boost efficiency. Remember to size the equipment correctly based on the building's load calculations, avoiding oversized systems that cycle on and off frequently. Correct sizing guarantees the equipment runs at its peak efficiency and minimizes energy waste. Always conduct thorough load calculations using relevant software and industry standards to determine the appropriate size of the equipment to meet the building's specific needs.

Beyond equipment choices, efficient operation is paramount. Scheduling your HVAC system to operate only when needed helps conserve energy. This is readily achieved through programmable thermostats that can be set to adjust temperatures based on occupancy patterns and time of day. Implementing smart thermostats with remote control capabilities can allow for greater flexibility and optimization based on real-time conditions and occupancy scheduling. Consider adjusting thermostat setpoints, even small changes can result in significant energy savings over time. Remember, every degree of temperature difference represents a tangible energy impact. For instance, increasing the thermostat setting in the summer by just 2 degrees can translate into substantial energy savings without a noticeable drop in comfort. Consistent maintenance is crucial. Regularly scheduled maintenance ensures the HVAC system functions at optimal efficiency. This involves tasks such as cleaning filters, checking refrigerant levels, and inspecting for any issues. Regular upkeep is far more cost-effective than emergency repairs and ensures the system's longevity and effective operation.

6.3 Renewable Energy Integration

Harnessing the sun's energy for HVAC systems offers significant environmental and economic benefits. Photovoltaic (PV) panels, directly converting sunlight into electricity, can power HVAC equipment, reducing reliance on the grid and lowering operational costs. The initial investment, while substantial, is often offset by long-term savings and government incentives, making it a financially attractive proposition for new constructions and retrofits. Careful consideration of panel placement, shading, and system sizing is paramount to maximize energy generation and system performance. Integration strategies should incorporate energy storage solutions, such as batteries, to address the intermittency of solar power, ensuring a reliable and consistent energy supply for HVAC operations, even during periods of low sunlight. This holistic approach considers both the immediate financial implications and the long-term sustainability of the integrated system. Proper system design and integration will prevent unexpected issues that could be costly to repair and could hinder the efficiency of the system.

Integrating solar thermal systems presents a viable alternative for heating applications. These systems employ solar collectors to absorb solar radiation, heating a fluid (e.g., water or glycol) that is subsequently used to provide heating for buildings. The heated fluid can be circulated directly to heating coils or used to preheat the air entering an air handling unit, significantly reducing the load on conventional heating equipment. System design depends greatly on factors such as climatic conditions, building orientation, and heating demand. The most crucial elements in solar thermal installations are proper sizing of the collector array and efficient heat storage and distribution. Improper sizing can significantly affect the performance of the HVAC system as a whole. It's crucial to analyze the building's specific heating requirements and its solar exposure to optimize the solar thermal system integration and achieve maximum efficiency.

Geothermal energy, drawing heat from the Earth's subsurface, provides a reliable and consistent source of energy for both heating and cooling applications. Geothermal heat pumps use the relatively stable ground temperature to transfer heat either into or out of a building, reducing reliance on fossil fuel-based heating and cooling systems. The upfront costs of installing a geothermal system are higher, involving the drilling of ground loops or the use of borehole heat exchangers. However, the long-term operational costs are typically lower, resulting in significant energy savings over the system's lifespan. The system's efficiency depends on several factors, including the ground's thermal properties, the depth of the ground loop, and the design of the heat exchanger. Careful geological surveys and accurate design are critical for successful geothermal integration, ensuring the system's efficient and reliable operation. Proper design and analysis will also reduce the amount of equipment failure and save the amount of energy used over the lifespan of the HVAC system.

Wind energy, while less directly applicable to HVAC systems compared to solar and geothermal options, can still contribute substantially to reducing the building's overall energy consumption. Small-scale wind turbines can generate electricity to power HVAC equipment, particularly advantageous in areas with consistently high wind speeds. However, considerations such as turbine size, noise levels, and visual impact on the surrounding environment must be carefully evaluated. The integration of wind energy requires careful site analysis to assess wind resource availability and to select appropriate turbine technology and placement. It is of great importance to understand that this option may not be suitable for all locations due to the inconsistent nature of wind and potential issues with noise pollution. Proper regulatory compliance, as well as community consultation, must be considered when planning to implement this kind of renewable energy into the HVAC system.

Biomass energy utilizes organic matter (e.g., wood chips, agricultural residues) to generate heat for building heating systems. This approach can be particularly sustainable in regions with readily available biomass resources. Biomass boilers or furnaces can be integrated into HVAC systems, providing a renewable source of heating energy. However, careful consideration must be given to the efficiency of combustion, emission controls, and ash management. The selection of appropriate biomass fuel and the design of the combustion system are crucial for minimizing environmental impact and maximizing efficiency. It's essential to consider the availability and cost of biomass fuel, storage, and disposal of ash when evaluating the overall sustainability of such a system. There are potential drawbacks when using biomass due to the inconsistencies of the energy it can provide.

The successful integration of renewable energy sources into HVAC systems hinges on careful planning and holistic system design. This includes a thorough energy audit, considering building-specific needs, available renewable energy resources, and potential cost-saving measures. The integration of advanced control systems is crucial to optimize the interaction between the renewable energy source and the HVAC equipment. System monitoring and data analysis also help identify opportunities for further optimization and ensure efficient operation. While the initial investment might seem high, the long-term savings in energy costs, coupled with the environmental benefits, make the integration of renewable energy a vital step towards sustainable and efficient building operations. By considering these factors and embracing a careful approach to design and implementation, building professionals can create HVAC systems that are efficient, cost-effective, and environmentally responsible.

6.4 Building Codes

Understanding building codes is paramount for any HVAC professional. Ignoring them can lead to costly fines, project delays, and, most importantly, safety hazards. These codes aren't arbitrary rules; they're based on years of research and real-world experience, designed to ensure efficient, safe, and comfortable building environments. Compliance is not merely a legal obligation, but a demonstration of professional responsibility and commitment to the well-being of building occupants. The consequences of non-compliance are far-reaching, impacting not only the project's immediate success but also its long-term sustainability and reputation. Therefore, a thorough understanding is essential for every HVAC project.

Local codes, however, vary significantly. What's permissible in one jurisdiction may be strictly prohibited in another. This geographic variability necessitates careful research and thorough understanding of the specific regulations governing each project location. For instance, energy efficiency standards vary dramatically across states and even municipalities, meaning your system design choices must be tailored to meet the specific requirements of the area. Failure to adequately research and understand local codes can result in project redesigns, extensive rework, and permit delays, significantly impacting project timelines and profitability. Diligent pre-design research is therefore essential.

Furthermore, the International Energy Conservation Code (IECC) plays a crucial role. While not a legally binding document in itself, it provides a benchmark and influences many local codes. Familiarization with the IECC's principles and recommendations can offer insights into prevailing best practices for energy efficiency and system design even if your locality doesn't strictly adhere to it. Understanding the IECC's structure and common requirements prepares you for the variations found in local building codes, enabling a faster and more effective adaptation process to the diverse requirements each project will present. The IECC's focus on energy efficiency also aligns with broader industry trends, making its study beneficial for staying ahead of regulatory changes.

Staying updated on code changes is equally important. Building codes are not static; they evolve to reflect technological advances and updated safety standards. Regularly checking for updates and revisions in your area is critical for successful project execution. Many jurisdictions provide online access to their building codes, allowing professionals to directly access the latest versions. Subscription to industry newsletters and participation in professional development courses can also contribute to staying informed and reducing the risks associated with outdated information. Continuous learning and staying informed are essential attributes of any successful HVAC professional.

Beyond the technical aspects, understanding the procedural side of building code compliance is crucial. Knowing how to navigate the permit application process, understanding the inspection requirements, and correctly documenting your design choices are essential steps to smooth project execution. Proactive communication with building inspectors ensures a clear understanding of the codes' requirements and avoids potential misunderstandings or disputes. A strong working relationship with local building authorities is invaluable in ensuring a seamless approval process. It's about building a collaborative relationship focused on safety and compliance.

Finally, remember that adhering to building codes isn't just about avoiding penalties; it's about building a reputation for excellence. A history of code compliance signals reliability and professionalism, fostering trust with clients and ensuring the long-term success of your business. It's about building a foundation of trust and credibility within your professional community. Consider this approach not just a legal necessity, but a crucial element of building professional reputation and business success. This translates to securing more projects, enhancing client trust, and ensuring the long-term viability of your professional practice.

6.5 Lifecycle Costs

Understanding lifecycle costs is crucial for making informed decisions about HVAC system selection and design. It's not just about the initial purchase price; long-term expenses significantly impact overall project viability and return on investment. Failure to account for these costs can lead to unexpected financial burdens and operational inefficiencies down the line. We need to consider the entire picture—the entire lifetime of the system.

Replacing a component prematurely due to inadequate initial selection or poor maintenance significantly increases the total cost. A seemingly inexpensive initial purchase might end up being more costly in the long run due to frequent repairs, early replacement, and higher energy consumption. Think of it like buying a cheaper car; the initial savings could easily be offset by increased maintenance and fuel costs over the vehicle's lifespan. Similarly, an HVAC system with lower initial costs, but inferior efficiency, will drain more energy over time, pushing up your utility bills far beyond the savings at purchase. Therefore, a thorough lifecycle cost analysis should be integral to your HVAC design process. This comprehensive assessment encompasses all expenses associated with a system throughout its operational life. It begins with the initial capital investment, which includes equipment purchase, installation, permits, and other upfront costs. However, it goes far beyond these initial expenses; you'll need to project operating costs which are often more significant than the capital cost over time. These include energy consumption, maintenance, repairs, and eventual replacement. This is where the true financial weight of your decision is revealed.

Operational costs, especially energy consumption, are significant contributors to total lifecycle costs. Energy efficiency is paramount; high-efficiency equipment may command a higher initial price, but the long-term energy savings often outweigh this initial investment. Factors like climate, building design, and occupancy patterns influence energy consumption, requiring careful consideration in the selection of HVAC systems. This emphasizes the importance of using accurate energy modeling software to predict energy performance and

to tailor your system to the specific conditions of the space.
Regular maintenance is also an essential component of lifecycle cost management. Preventative maintenance prevents more costly repairs, while neglecting this results in premature component failure and unexpected downtime. A schedule of regular inspections, filter changes, and cleaning will keep your system running efficiently and extend its lifespan. This means investing in regular inspections and maintenance plans, which, although might seem like an additional expense, is crucial for minimizing costly repairs. This is an investment that pays off in the long run. Consider the cost of downtime as well; a faulty HVAC system can lead to productivity losses, discomfort, and even damage to the building and its contents.
Replacement costs are inevitable but can be mitigated through proper planning and system selection. The lifespan of HVAC equipment varies considerably depending on quality, maintenance, and operating conditions. It's important to factor in the anticipated replacement cost when comparing different system options. This will assist you in forming a balanced and realistic picture of the cost-effectiveness of your choice. Consider that the price of equipment increases over time, so you must account for this potential inflation when forecasting fu-

ture replacement costs.

In summary, a holistic lifecycle cost analysis is a powerful tool in the HVAC system selection process. It encourages a long-term perspective, moves beyond the initial purchase price, and promotes making informed decisions based on total ownership cost. This analysis should incorporate the initial capital outlay, operating costs, maintenance expenses, potential repair costs and, finally, the eventual replacement costs of the equipment. By conducting this analysis carefully, you're setting yourself up for a successful HVAC system implementation, ensuring optimal performance and minimizing long-term expenses. Remember, a small initial saving can be easily eroded by significantly higher operating or repair costs in the long run. Always consider the whole picture.

7 Advanced HVAC Design

Designing sophisticated HVAC systems necessitates a deeper understanding beyond fundamental principles. We move beyond simple calculations and delve into the realm of advanced modeling and simulation. This allows for the prediction of system performance under various operating conditions, optimizing design choices before implementation and minimizing costly errors down the line. Sophisticated software packages, employing computational fluid dynamics (CFD) and other numerical techniques, allow engineers to visualize airflow patterns, temperature distributions, and pressure drops within complex building geometries. This detailed level of analysis enables the identification and mitigation of potential issues, such as uneven air distribution or excessive energy consumption, leading to more efficient and effective systems. Accurate modeling also aids in the selection of appropriate equipment sizes, reducing both capital costs and operational expenses.

The design process extends far beyond standard configurations. Specialized applications demand tailored solutions. Consider cleanrooms, for instance, where stringent control over temperature, humidity, and particle counts is paramount. These environments require highly specialized filtration systems, precise airflow management, and often, the implementation of dedicated pressure control strategies to maintain a sterile atmosphere and prevent contamination. Similarly, large-scale systems, such as those found in district cooling networks, present unique challenges related to energy distribution and efficient management of resources across multiple buildings. These projects demand in-depth understanding of network dynamics, energy storage, and optimization techniques to balance supply and demand effectively and economically.

Another area where advanced design knowledge is crucial is in the integration of renewable energy sources. While previous chapters touched on energy efficiency, this chapter highlights the sophisticated strategies used to incorporate renewable sources effectively into HVAC systems. This can involve integrating solar thermal collectors for water heating, harnessing geothermal energy for heating and cooling, or utilizing waste heat recovery systems to improve overall efficiency. These integrations require careful consideration of system compatibility, energy storage needs, and the development of intelligent control strategies to optimize energy utilization. Advanced HVAC design therefore necessitates understanding of complex energy interactions and innovative technological solutions to maximize renewable energy contribution.

Let's consider hospitals, a critical application where maintaining precise environmental control is not merely a matter of comfort, but directly impacts patient health and recovery. Hospital HVAC systems require rigorous attention to infection control, often integrating sophisticated air filtration systems such as HEPA (High-Efficiency Particulate Air) filters to eliminate airborne pathogens. These systems demand detailed analysis to prevent cross-contamination between patient areas, ensuring an optimal and safe environment. Careful zone control is another key aspect, allowing for independent regulation of temperature and air quality in different areas of the hospital, such as operating rooms, patient rooms, and waiting areas. This level of control necessitates advanced control systems and robust monitoring capabilities.

Beyond hospitals, consider data centers – critical infrastructure for the modern digital world. These facilities house countless servers that generate significant heat, demanding efficient cooling strategies. Advanced HVAC designs for data centers often incorporate advanced cooling technologies, such as liquid cooling systems or specialized air handling units, to effectively manage heat loads and ensure optimal operating conditions for sensitive equipment. Redundancy is another key element in data center HVAC design, ensuring uninterrupted operation even in the event of equipment failure. This involves designing systems with backup components and implementing intelligent control strategies that can switch to redundant systems seamlessly. Furthermore, data centers are often located in environments with extreme climate conditions, requiring consideration of factors like external temperature variations and humidity levels in the design process.

Finally, let's examine the importance of life-cycle analysis in advanced HVAC design. It goes beyond immediate installation costs. This comprehensive approach evaluates all costs associated with the system throughout its entire lifespan, including initial investment, operating costs, maintenance expenses, and eventual replacement. A detailed life-cycle analysis allows for informed decision-making regarding equipment selection, system design, and operational strategies, maximizing long-term cost-effectiveness. This evaluation should consider factors such as energy efficiency, maintenance requirements, and the system's expected lifespan, ensuring that the chosen design represents the most economically viable and environmentally responsible solution over its operating lifetime. This holistic approach necessitates considering factors beyond initial cost, enabling informed decisions that promote both sustainability and financial prudence.

7.1 Modeling and Simulation

Modeling and simulation are invaluable tools in modern HVAC design. They allow engineers to virtually test and refine systems before physical construction, minimizing costly errors and ensuring optimal performance. Unlike relying solely on empirical data or simplified calculations, sophisticated software allows for a detailed analysis of complex interactions between components. This leads to more efficient systems and significant cost savings in the long run.

For instance, imagine designing a large-scale HVAC system for a new hospital. The sheer number of interconnected zones, varying loads, and precise temperature requirements would make manual calculations incredibly complex and potentially inaccurate. Simulation software enables engineers to input all relevant parameters – from building geometry and insulation values to occupancy patterns and equipment specifications – to predict system behavior under diverse conditions. This includes not only steady-state performance but also transient responses to changing external factors like weather conditions. The results provide crucial insights into energy consumption, peak loads, and potential system bottlenecks, guiding design optimization before any construction begins.

This detailed modeling also permits a "what-if" scenario analysis. What happens if we change the air handler location? What is the impact of using a different refrigerant? How does increased insulation affect energy usage over the entire building lifecycle? Such analyses are impossible to perform practically without resorting to computationally intensive simulations. Through the systematic variation of parameters and observation of the resultant effects, engineers can find the most effective and energy-efficient design.

Beyond simply predicting performance, simulation helps in optimizing control strategies. By integrating control algorithms within the model, engineers can test different strategies to determine which best meets the specified comfort levels and energy targets. This could entail examining various PID controller settings, comparing different scheduling approaches, or exploring the impact of advanced control methods like model predictive control. The virtual environment allows for experimentation without any risk to the actual system.

However, accurate simulation depends critically on the quality of input data. Incomplete or inaccurate building information, incorrect equipment specifications, or overly simplified models can produce misleading results. Careful attention to data accuracy is paramount. You must verify all inputs meticulously, ensuring they reflect real-world conditions. The use of high-quality building information modeling (BIM) data can significantly enhance the accuracy and reliability of simulations.

Furthermore, validation is essential. Once a simulation model is built and analyzed, it must be validated against real-world data. If possible, comparing simulation results to performance data from a similar, already-constructed system provides a valuable benchmark. This iterative process of model refinement based on real-world observations ensures that simulations remain a useful tool, continuously improving their accuracy and relevance.

Therefore, embracing modeling and simulation is not merely an advanced technique; it's a crucial step toward designing efficient, reliable, and cost-effective HVAC systems. It empowers engineers to make informed decisions, minimize risks, and create innovative solutions that meet the demands of increasingly complex building environments. The ability to explore various options before investing substantial resources in construction represents a significant advantage. And while the initial learning curve may seem steep, the long-term benefits far outweigh the initial investment of time and resources. Mastering these techniques is essential for any professional seeking excellence in HVAC system design. The transition from basic calculations to comprehensive modeling marks a critical shift towards precision and sophisticated HVAC system engineering. Modern practice requires the use of simulation to meet industry demands and environmental standards. It offers a future-forward approach to a field constantly evolving to meet new demands and improve existing technologies. Thus, incorporating it into your skillset can only enhance your standing within the field and contribute to more effective and sustainable solutions. The benefits extend beyond mere cost-effectiveness, ultimately contributing to more environmentally friendly and efficient HVAC systems that benefit both building owners and occupants alike.

7.2 Special System Designs

Designing for unique needs presents exciting challenges. Consider a hospital operating room; maintaining a sterile, precisely controlled environment requires a system far beyond a standard HVAC setup. We're talking HEPA filtration, laminar airflow, and potentially even positive pressure to prevent the entry of contaminants. The design must account for the higher heat loads generated by equipment and personnel, the stringent hygiene requirements necessitating specialized materials and cleaning protocols, and the potential for complex interactions between the HVAC system and medical devices. This isn't simply about temperature regulation; it's about actively safeguarding the health and safety of patients. Careful consideration of every component, from the air handlers to the final diffusers, is paramount.

Cleanrooms, similarly, demand exacting standards. These environments, crucial for industries like pharmaceuticals and semiconductor manufacturing, require incredibly precise control over temperature, humidity, and particulate matter. Here, the focus shifts to minimizing particle generation and ensuring the removal of even microscopic contaminants. The design involves specialized filters, careful airflow management to create unidirectional or laminar flow, and often, sophisticated monitoring systems to track environmental parameters in real-time. This necessitates a deep understanding of airflow dynamics, filter technology, and the unique challenges of each application. The goal is not just comfort, but the prevention of product contamination and process failure, potentially costing millions.

Data centers present a different set of complexities. These facilities house vast amounts of sensitive electronic equipment, which generates significant heat. Efficient cooling is crucial to preventing equipment failure and data loss. Specialized systems, often utilizing liquid cooling or precision air conditioning, are employed to maintain optimal operating temperatures within tight tolerances. Redundancy and fault tolerance are often built into the design to ensure uninterrupted operation, which adds to the overall complexity. Designing these systems demands a thorough understanding of heat dissipation mechanisms, fluid dynamics, and the specific thermal requirements of the equipment involved. Effective solutions prioritize operational continuity alongside energy efficiency, maximizing performance while minimizing operational cost.

Then there are systems designed for unique climates. In extremely hot and humid climates, specialized dehumidification techniques may be necessary to maintain comfortable indoor conditions. This could involve employing desiccant dehumidifiers, which work by absorbing moisture from the air, alongside traditional refrigeration-based methods. In colder climates, efficient heating and snow-melt systems might be integral parts of the overall design, minimizing energy loss and ensuring safe access to buildings. Adaptation for extreme weather conditions requires careful consideration of local climatic data and a solid understanding of the performance characteristics of different HVAC components under stress. Resilience and reliability become top priorities.

Consider also systems integrated with renewable energy sources. The integration of solar thermal collectors, geothermal heat pumps, or wind turbines can significantly reduce reliance on fossil fuels and decrease the environmental impact of the building. Designing these systems, however, requires specialized knowledge of renewable energy technologies and their integration with existing HVAC infrastructure. This often includes careful sizing of components, optimization of energy storage solutions, and understanding the dynamic interaction between renewable energy generation and building loads. The goal here is a balanced, sustainable system that leverages both efficient conventional components and cleaner, renewable technologies.

Finally, understanding and managing the interactions between different subsystems is crucial in special system designs. For instance, the HVAC system might need to coordinate with the building's fire suppression system, ensuring that critical safety features aren't compromised during operation or in the event of an emergency. Integrating energy management systems that control lighting, occupancy sensors, and other building functions can further optimize energy use. Thorough analysis of these interactions and careful system integration are essential for ensuring smooth operation and avoiding conflicts. The overall goal should be to ensure a harmonized, safe, and efficient environment. This is more than mere engineering; it's about building a resilient and sustainable future.

www.ingramcontent.com/pod-product-compliance
Lightning Source LLC
LaVergne TN
LVHW020234210926
848481LV00016B/2206
* 9 7 9 8 3 4 8 4 9 7 4 8 4 *